LONDON MATHEMATICAL SOCIETY LECTURE NOTE SERIES

Managing Editor: Professor I.M. James,
Mathematical Institute, 24-29 St Giles, Oxford

1. General cohomology theory and K-theory, P.HILTON
4. Algebraic topology, J.F.ADAMS
5. Commutative algebra, J.T.KNIGHT
8. Integration and harmonic analysis on compact groups, R.E.EDWARDS
9. Elliptic functions and elliptic curves, P.DU VAL
10. Numerical ranges II, F.F.BONSALL & J.DUNCAN
11. New developments in topology, G.SEGAL (ed.)
12. Symposium on complex analysis, Canterbury, 1973, J.CLUNIE
    & W.K.HAYMAN (eds.)
13. Combinatorics: Proceedings of the British Combinatorial Conference
    1973, T.P.McDONOUGH & V.C.MAVRON (eds.)
15. An introduction to topological groups, P.J.HIGGINS
16. Topics in finite groups, T.M.GAGEN
17. Differential germs and catastrophes, Th.BROCKER & L.LANDER
18. A geometric approach to homology theory, S.BUONCRISTIANO, C.P. BOURKE
    & B.J.SANDERSON
20. Sheaf theory, B.R.TENNISON
21. Automatic continuity of linear operators, A.M.SINCLAIR
23. Parallelisms of complete designs, P.J.CAMERON
24. The topology of Stiefel manifolds, I.M.JAMES
25. Lie groups and compact groups, J.F.PRICE
26. Transformation groups: Proceedings of the conference in the University
    of Newcastle-upon-Tyne, August 1976, C.KOSNIOWSKI
27. Skew field constructions, P.M.COHN
28. Brownian motion, Hardy spaces and bounded mean oscillations,
    K.E.PETERSEN
29. Pontryagin duality and the structure of locally compact Abelian
    groups, S.A.MORRIS
30. Interaction models, N.L.BIGGS
31. Continuous crossed products and type III von Neumann algebras,
    A.VAN DAELE
32. Uniform algebras and Jensen measures, T.W.GAMELIN
33. Permutation groups and combinatorial structures, N.L.BIGGS & A.T.WHITE
34. Representation theory of Lie groups, M.F. ATIYAH et al.
35. Trace ideals and their applications, B.SIMON
36. Homological group theory, C.T.C.WALL (ed.)
37. Partially ordered rings and semi-algebraic geometry, G.W.BRUMFIEL
38. Surveys in combinatorics, B.BOLLOBAS (ed.)
39. Affine sets and affine groups, D.G.NORTHCOTT
40. Introduction to Hp spaces, P.J.KOOSIS
41. Theory and applications of Hopf bifurcation, B.D.HASSARD,
    N.D.KAZARINOFF & Y-H.WAN
42. Topics in the theory of group presentations, D.L.JOHNSON
43. Graphs, codes and designs, P.J.CAMERON & J.H.VAN LINT
44. Z/2-homotopy theory, M.C.CRABB
45. Recursion theory: its generalisations and applications, F.R.DRAKE
    & S.S.WAINER (eds.)
46. p-adic analysis: a short course on recent work, N.KOBLITZ
47. Coding the Universe, A.BELLER, R.JENSEN & P.WELCH
48. Low-dimensional topology, R.BROWN & T.L.THICKSTUN (eds.)

49. Finite geometries and designs, P.CAMERON, J.W.P.HIRSCHFELD & D.R.HUGHES (eds.)
50. Commutator calculus and groups of homotopy classes, H.J.BAUES
51. Synthetic differential geometry, A.KOCK
52. Combinatorics, H.N.V.TEMPERLEY (ed.)
53. Singularity theory, V.I.ARNOLD
54. Markov processes and related problems of analysis, E.B.DYNKIN
55. Ordered permutation groups, A.M.W.GLASS
56. Journées arithmétiques 1980, J.V.ARMITAGE (ed.)
57. Techniques of geometric topology, R.A.FENN
58. Singularities of smooth functions and maps, J.MARTINET
59. Applicable differential geometry, F.A.E.PIRANI & M.CRAMPIN
60. Integrable systems, S.P.NOVIKOV et al.
61. The core model, A.DODD
62. Economics for mathematicians, J.W.S.CASSELS
63. Continuous semigroups in Banach algebras, A.M.SINCLAIR
64. Basic concepts of enriched category theory, G.M.KELLY
65. Several complex variables and complex manifolds I, M.J.FIELD
66. Several complex variables and complex manifolds II, M.J.FIELD
67. Classification problems in ergodic theory, W.PARRY & S.TUNCEL
68. Complex algebraic surfaces, A.BEAUVILLE
69. Representation theory, I.M.GELFAND et. al.
70. Stochastic differential equations on manifolds, K.D.ELWORTHY
71. Groups - St Andrews 1981, C.M.CAMPBELL & E.F.ROBERTSON (eds.)
72. Commutative algebra: Durham 1981, R.Y.SHARP (ed.)
73. Riemann surfaces: a view toward several complex variables, A.T.HUCKLEBERRY
74. Symmetric designs: an algebraic approach, E.S.LANDER
75. New geometric splittings of classical knots (algebraic knots), L.SIEBENMANN & F.BONAHON
76. Linear differential operators, H.O.CORDES
77. Isolated singular points on complete intersections, E.J.N.LOOIJENGA
78. A primer on Riemann surfaces, A.F.BEARDON
79. Probability, statistics and analysis, J.F.C.KINGMAN & G.E.H.REUTER (eds.)
80. Introduction to the representation theory of compact and locally compact groups, A.ROBERT
81. Skew fields, P.K.DRAXL

London Mathematical Society Lecture Note Series. 81

Skew Fields

P.K. DRAXL
Reader in Mathematics, University of Bielefeld

CAMBRIDGE UNIVERSITY PRESS
Cambridge
London New York New Rochelle
Melbourne Sydney

CAMBRIDGE UNIVERSITY PRESS
Cambridge, New York, Melbourne, Madrid, Cape Town, Singapore, São Paulo

Cambridge University Press
The Edinburgh Building, Cambridge CB2 8RU, UK

Published in the United States of America by Cambridge University Press, New York

www.cambridge.org
Information on this title: www.cambridge.org/9780521272742

© Cambridge University Press 1983

This publication is in copyright. Subject to statutory exception
and to the provisions of relevant collective licensing agreements,
no reproduction of any part may take place without the written
permission of Cambridge University Press.

First published 1983
Re-issued in this digitally printed version 2007

A catalogue record for this publication is available from the British Library

Library of Congress Catalogue Card Number: 82-22036

ISBN 978-0-521-27274-2 paperback

Max Deuring

zum 75. Geburtstag

am 9. Dezember 1982

gewidmet

CONTENTS

		page
Preface		vii
Conventions on Terminology		ix
Part I.	Skew Fields and Simple Rings	1
§ 1.	Some ad hoc Results on Skew Fields	3
§ 2.	Rings of Matrices over Skew Fields	8
§ 3.	Simple Rings and Wedderburn's Main Theorem	13
§ 4.	A Short Cut to Tensor Products	18
§ 5.	Tensor Products and Algebras	25
§ 6.	Tensor Products and Galois Theory	33
§ 7.	Skolem-Noether Theorem and Centralizer Theorem	39
§ 8.	The Corestriction of Algebras	50
Part II.	Skew Fields and Brauer Groups	57
§ 9.	Brauer Groups over Fields	59
§ 10.	Cyclic Algebras	71
§ 11.	Power Norm Residue Algebras	77
§ 12.	Brauer Groups and Galois Cohomology	92
§ 13.	The Formalism of Crossed Products	97
§ 14.	Quaternion Algebras	103
§ 15.	p-Algebras	106
§ 16.	Skew Fields with Involution	112
§ 17.	Brauer Groups and K_2-Theory of Fields	119
§ 18.	A Survey of some further Results	122
Part III.	Reduced K_1-Theory of Skew Fields	125
§ 19.	The Bruhat Normal Form	127
§ 20.	The Dieudonné Determinant	133
§ 21.	The Structure of $SL_n(D)$ for $n \geq 2$	140

§ 22.	Reduced Norms and Traces	143
§ 23.	The Reduced Whitehead Group $SK_1(D)$ and Wang's Theorem	155
§ 24.	$SK_1(D) \neq 1$ for suitable D	167
§ 25.	Remarks on $USK_1(D,I)$	171
Bibliography		173
Thesaurus		179
Index		180

PREFACE

This is a substantially extended version of the notes on 25 lectures delivered at the Pennsylvania State University during Spring Term 1981 under 572.2 ("Special Topics in Algebra").

Most of the material has been presented earlier elsewhere: in the Seminar Bielefeld-Göttingen during the Summer Term 1967 (mainly Part III), in lectures at the Universität Bielefeld during the Academic Year 1977/78 and in lectures at the Université de Grenoble in February/March 1979 (mainly §§19/20.). Some of the material has been discussed afterwards in the course of different lectures which I delivered at the Universität Bielefeld during the Academic Year 1981/82.

The text falls into three parts: Skew fields and simple rings, Skew fields and Brauer groups, and Reduced K_1-theory of skew fields. As regards their contents the reader is advised to consult the introductory remarks at the beginning of each of these parts on pages 1, 57 and 125 respectively.

During the preparation of the final draft of these notes I have enjoyed the assistance of B. Fein (Oregon State University), D. Garbe (Universität Bielefeld), I.M. James (Oxford University, editor of the LMS Lecture Note Series), Ch. Preston (Universität Bielefeld), S. Rosset (Tel-Aviv University), B. Weisfeiler (Pennsylvania State University) and J. Tate (Harvard University, who communicated so far unpublished work to me and gave permission to publish his arguments here; cf. §11.); I am grateful for their help.

Moreover, I want to take this opportunity to express gratitude to G. Andrews, D. Brownawell, D. James, D. Rung, L.N. Vaserstein and B. Weisfeiler who enabled my visit to the Pennsylvania State University (during the Academic Year 1980/81) or made the stay there so enjoyable for me and my family.

Finally I must report that I have greatly benefitted from the help of P.M. Cohn (Bedford College, University of London) whose assistance and encouragement only made it possible for this book to appear. I am most grateful for all his help.

       Bielefeld
       August 1982

CONVENTIONS ON TERMINOLOGY

As usual, $\mathbb{N}, \mathbb{Z}, \mathbb{Q}, \mathbb{R}, \mathbb{C}$ stand for the natural numbers, integers, rational numbers, real numbers and complex numbers respectively. The French reader should note $0 \notin \mathbb{N}$.

We write $X := Y$ if X is defined by Y.

From group theory we adopt the following (standard) notation:

$H \leq G$	H is a subgroup of G,
$H \trianglelefteq G$	H is a normal subgroup of G,
$[G,G]$	commutator subgroup of G,
$G^{ab} := G/[G,G]$	commutator factor group of G,
$Z(G)$	centre of G.

If R is a ring, then we usually assume that it has a unit element, denoted by 1 (not necessarily $1 \neq 0$; note that $1 = 0$ implies $R = \{0\}$), which is inherited by subrings, preserved by homomorphisms and acts unitally on all R-modules. Moreover, we use the notation:

R^+	additive group of R,
R^*	multiplicative group of R (if $1 \neq 0$),
$Z(R)$	centre of R,
$M_n(R)$	ring of (n,n)-matrices over R,
$GL_n(R) := M_n(R)^*$	general linear group over R.

We call a ring D a *skew field* if $D^* \cup \{0\} = D$. We assume the reader to be familiar with the fact that left/right vector spaces over a skew field are always free of unique rank (called the *left/right dimension*); the proofs of these facts work precisely as in the commutative case (known from Linear Algebra).

Finally we point out that we assume a good knowledge of (ordinary) Field Theory (including Galois Theory of finite field extensions); here *field* stands for commutative field and F_q stands for the (finite) field with q elements.

PART I . SKEW FIELDS AND SIMPLE RINGS

The history of skew fields begins with quaternions, whose discovery W.R. Hamilton (1805-1865) regarded as the climax of his career. F. Klein [1926/27,p.184 in vol.1] writes in his famous treatise "Vorlesungen über die Entwicklung der Mathematik im 19. Jahrhundert" (which is an outstanding account):

...

Von hier aus entwickelte sich nun bei Hamilton das größte Interesse an der Fragestellung, ob man die nützliche, geometrische Interpretation des Rechnens mit $x + iy$ in der Ebene nicht irgendwie - durch Schaffung neuer komplexer Zahlen - auf den Raum, d.h. unsern gewöhnlichen R_3, übertragen könne. Seine unermüdlichen Anstrengungen führen ihn endlich 1843 zur Erfindung der *Quaternionen*, d.h. geeigneter viergliedriger Zahlen, deren Erforschung und Verbreitung er sich fortan ausschließlich widmete. Ihre Theorie legte er dar in den beiden ausführlichen Werken:

Lectures on Quaternions, Dublin 1853

Elements on Quaternions, London 1866 (posthum).

Sehr bald wurden die Quaternionen in Dublin ein alles andere überragender Gegenstand des mathematischen Interesses, ja sogar ein offizielles Examensfach, ohne dessen Kenntnis keine Absolvierung des College mehr denkbar war. Hamilton selbst gestaltete sie für sich zu einer Art orthodoxer Lehre des mathematischen Credo, in die er alle seine geometrischen und sonstigen Interessen hineinzwang, je mehr sich gegen Ende seines Lebens sein Geist vereinseitigte und unter den Folgen des Alkohols verdüsterte.

...

In Part I of these lectures we start with a brief description of Hamilton's quaternions; however, we do not take his point of view since we use a definition involving matrices and these were only later introduced into mathematics by A. Cayley (1821-1895) in 1855. All this is done in §1. which also includes some remarks on (skew) formal Laurent series fields introduced by D. Hilbert (1862-1943) in 1898. (Later, in Part II. (§ 14.) we shall come back to the quaternions from a more abstract point of view.) In §§2/3 we develop a theory of simple rings as suggested by E. Artin (1898-1962) in the late 1920's; important special cases of the material presented here have been introduced by J.H.M. Wedderburn (1882--1948) as early as 1907. In §§4/5/6 we discuss certain techniques involving tensor products which are relevant to our subject. §7. contains the backbone of these lectures, the Skolem-Noether Theorem, proved in 1927 by T. Skolem (1887-1963) and rediscovered in 1933 by E. Noether (1882--1935); we treat this theorem in a setting which goes back to E. Artin and G. Whaples (1914-1981). We close Part I with a discussion of the corestriction of algebras intruduced by C. Riehm [1970]; here (in §8.) we present only a simplified version which is sufficient for our purposes.

Roughly speaking one may say that §§2,..,7 comprise a slightly modified and modernized version of the first seven chapters of the classical set of notes by E. Artin *et al.* [1948]; however, in our lectures we do not discuss (and make no use of) semisimple rings; those interested in such things may consult for example C.W. Curtis & I. Reiner [1962] or vol. 2 of P.M. Cohn [1974/77] (cf. also some of the exercises).

§ 1. SOME AD HOC RESULTS ON SKEW FIELDS

Consider the set of matrices

(1) $$H := \left\{ \begin{pmatrix} z & u \\ -\bar{u} & \bar{z} \end{pmatrix} \,\bigg|\, z, u \in \mathbb{C} \right\} \subseteq M_2(\mathbb{C})$$

where $\bar{z}$ denotes the complex conjugate of $z \in \mathbb{C}$. An easy calculation shows that H is in fact a ring with unit element the unit matrix 1. If $\begin{pmatrix} z & u \\ -\bar{u} & \bar{z} \end{pmatrix} \neq 0$, i.e. $|z|^2 + |u|^2 \neq 0$, then

$$\begin{pmatrix} z & u \\ -\bar{u} & \bar{z} \end{pmatrix}^{-1} = (|z|^2 + |u|^2)^{-1} \begin{pmatrix} \bar{z} & -u \\ \bar{u} & z \end{pmatrix} \in H \;,$$

hence H is even a skew field, called the skew field of (*ordinary* or *real*) *quaternions*. Of course, H is a 4-dimensional R-vector space with basis

$$1 = \begin{pmatrix} 1 & 0 \\ 0 & 1 \end{pmatrix}, \; i := \begin{pmatrix} i & 0 \\ 0 & -i \end{pmatrix}, \; j := \begin{pmatrix} 0 & 1 \\ -1 & 0 \end{pmatrix}, \; k := \begin{pmatrix} 0 & i \\ i & 0 \end{pmatrix}$$

with the usual $i \in \mathbb{C}$ satisfying $i^2 = -1$. The elements $1, i, j, k$ satisfy the multiplication table on the right. Usually one writes $a1 + bi + cj + dk$ in place of $\begin{pmatrix} a+bi & c+di \\ -c+di & a-bi \end{pmatrix}$ ($a,b,c,d \in \mathbb{R}$).

	1	i	j	k
1				
i		-1	k	$-j$
j		$-k$	-1	i
k		j	$-i$	-1

Obviously the eight elements $1, i, j, k, -1, -i, -j, -k \in H^*$ form a finite subgroup of the multiplicative group H^* of our skew field H - called the *Quaternion Group* - which clearly is non commutative and hence not a cyclic group; this could never happen in a commutative field since any finite subgroup of the multiplicative group of a commutative field is necessarily cyclic (cf. Field Theory). Moreover the equation $x^2 + 1 = 0$ obviously has the six solutions $i, j, k, -i, -j, -k \in H$; over a commutative field it could have at most two solutions. Here the above equation even has an infinite number of solutions: choose $b,c,d \in R$ such that $b^2 + c^2 + d^2 = 1$, then a straightforward calculation shows $(bi + cj + dk)^2 = \ldots = -1$. This phenomenon will be understood later (cf. Example 1 in §7.).

Now consider the injection $R \to H$, $t \mapsto t1$; this makes R a commutative subfield of the skew field H such that $R \subseteq Z(H)$, but even $R = Z(H)$ holds:
indeed, assume $\begin{pmatrix} z & u \\ -\bar{u} & \bar{z} \end{pmatrix} \in Z(H)$, hence $\begin{pmatrix} z & u \\ -\bar{u} & \bar{z} \end{pmatrix} \begin{pmatrix} 0 & v \\ -\bar{v} & 0 \end{pmatrix} =$
$= \begin{pmatrix} 0 & v \\ -\bar{v} & 0 \end{pmatrix} \begin{pmatrix} z & u \\ -\bar{u} & \bar{z} \end{pmatrix}$ for all $v \in C$. This amounts to $vz = v\bar{z}$ and $u\bar{v} = \bar{u}v$ for all $v \in C$, hence $z \in R$ and $u = 0$. Therefore we have $|H:Z(H)| = |H:R| = 4 = 2^2$. The fact that $Z(H)$ is a field is not surprising, in fact we have

Lemma 1. *Let D be a skew field, then $Z(D)$ is a commutative subfield of D.*

Proof. Obviously $Z(D)$ is a subring and even a field since $zd = dz$ for all $d \in D$ clearly implies $d^{-1}z^{-1} = z^{-1}d^{-1}$ for all $d \in D$ ($z,d \neq 0$) for any given $z \in Z(D)$. □

The above example may be generalized as follows: replace the extension C/R by an arbitrary separable quadratic extension L/K, select some $a \in K^*$ and consider the set of matrices

(2) $\quad D := D_a(L/K) := \left\{ \begin{pmatrix} z & u \\ a\bar{u} & \bar{z} \end{pmatrix} \bigg| z,u \in L \right\} \subseteq M_2(L)$

where $\bar{z}$ denotes the conjugate of $z \in L$. Again D is a ring, and an easy calculation shows that

$\begin{pmatrix} z & u \\ a\bar{u} & \bar{z} \end{pmatrix}^{-1}$ exists if and only if $z\bar{z} - au\bar{u} \neq 0$, i.e. if

and only if a is *not* a norm for the extension L/K. Moreover we have the formula

$\begin{pmatrix} z & u \\ a\bar{u} & \bar{z} \end{pmatrix}^{-1} = (z\bar{z}-au\bar{u})^{-1} \begin{pmatrix} \bar{z} & -u \\ -a\bar{u} & z \end{pmatrix}$,

provided either side exists. Again D is a 4-dimensional K-algebra with $Z(D) = K$ (here, as above, we identify $t \in K$ with $t1 \in D$); this follows in the same way as in the case of the real quaternions.

Summarizing our remarks gives

Lemma 2. $D = D_a(L/K)$ according to (2) *is a 4-dimensional K-algebra with centre* K, *and it is a skew field if and only if* a *is not a norm for the extension* L/K. □

Let us now study another classical example: let L be a commutative field and σ an automorphism of L. Call $K := \text{Fix}_L(\sigma) := \{ x \in L \mid \sigma(x) = x \}$ the fixed field of σ in L.

Definition 1. *Denote by* $L((T;\sigma))$ *the ring of formal Laurent series* $\sum_{i=R}^{\infty} a_i T^i$ *in the indeterminate* T *with coefficients* $a_i \in L$ ($R \in \mathbb{Z}$), *with usual addition but skew multiplication such that* $Ta = \sigma(a)T$, *i.e.* $T^i a = \sigma^i(a) T^i$ ($a \in L$).

If $\sigma = \text{id}_L$ then $L((T;\sigma))$ is the usual commutative field of formal Laurent series over L in T, customarily denoted $L((T))$. We want to show that $L((T;\sigma))$ is always a skew field: indeed, let $0 \neq d = \sum_{i=R}^{\infty} a_i T^i$ ($a_R \neq 0$) be given, let us calculate its inverse

$$d^{-1} = \sum_{j=S}^{\infty} x_j T^j \quad (x_S \neq 0)$$

where $S \in \mathbb{Z}$ and the $x_j \in L$ are not yet known. We have necessarily

$$T^0 = 1 = (\sum_{j=S}^{\infty} x_j T^j)(\sum_{i=R}^{\infty} a_i T^i) =$$

$$= \sum_{r=R+S}^{\infty} (\sum_{i+j=r} x_j \sigma^j(a_i))T^r ,$$

hence $R + S = 0$, i.e. $S = -R$. Comparing coefficients at $r = 0$ gives

(3) $\quad x_{-R} = \sigma^{-R}(a_R^{-1}) \neq 0$;

doing the same for $r \geq 1$ we get immediately

$$0 = \sum_{i+j=r} x_j \sigma^j(a_i) = \sum_{j=-R}^{-R+r} x_j \sigma^j(a_{r-j}) ,$$

i.e. the recurrence relations

(4) $\quad x_{-R+r} = - \sigma^{-R+r}(a_R^{-1}) \sum_{j=-R}^{-R+r-1} x_j \sigma^j(a_{r-j})$ ($r \geq 1$).

Hence for $d \neq 0$ we may calculate the coefficients x_j of its inverse d^{-1} successively with the aid of (3) and (4). Therefore we have proved

Lemma 3. *The ring* $L((T;\sigma))$ *according to Definition 1 is a skew field.* □

Now let L again be a commutative field (cf. Lemma 2 in §24.), then:

Lemma 4. *Let* $D := L((T;\sigma))$ *be given. If σ has infinite order, then* $Z(D) = K$, *hence* $|D:Z(D)| = \infty$; *if σ has the finite order n in* $\mathrm{Aut}(L)$, *then* $Z(D) = K((T^n))$, *hence* $|D:Z(D)| = n^2$ ($K = \mathrm{Fix}_L(\sigma)$).

Proof. "$\supseteq$" is obvious in both cases; let us prove the converse: pick an element
$$z = \sum_{i=R}^{\infty} z_i T^i \in Z(D) \text{ , i.e. } az = za \text{ for all } a \in D \text{ .}$$

Set $a = \sum_{j=S}^{\infty} a_j T^j$ ($a_S \neq 0$, because we may assume $a \neq 0$) , then $az = za$ amounts to

$$\sum_{r=R+S}^{\infty} (\sum_{i+j=r} a_j \sigma^j(z_i)) T^r = (\sum_{j=S}^{\infty} a_j T^j)(\sum_{i=R}^{\infty} z_i T^i) = az = za =$$
$$= (\sum_{i=R}^{\infty} z_i T^i)(\sum_{j=S}^{\infty} a_j T^j) = \sum_{r=R+S}^{\infty} (\sum_{i+j=r} z_i \sigma^i(a_j)) T^r \text{ ,}$$

hence

(5) $\quad \sum_{j=S}^{-R+r} z_{r-j} \sigma^{r-j}(a_j) = \sum_{j=S}^{-R+r} a_j \sigma^j(z_{r-j}) \quad$ for all $r \geq R + S$.

Now take in (5) $S := 1$, $a_1 := 1$ and $a_j := 0$ for $j \geq 2$. It follows $z_{r-1} = \sigma(z_{r-1})$ for all $r-1$, hence $z_r \in \mathrm{Fix}_L(\sigma) = K$ for all r , therefore
$$Z(D) \subseteq K((T)) \text{ in any case.}$$

Let us now suppose that σ has infinite order in $\mathrm{Aut}(L)$. Take in (5) $S := 1$, $a_1 \in L$ such that $\sigma^{r-1}(a_1) \neq a_1$ for any $r \neq 1$, and $a_j := 0$ for $j \geq 2$. It follows that $z_{r-1} \sigma^{r-1}(a_1) = a_1 \sigma(z_{r-1}) = a_1 z_{r-1}$; by construction of a_1 this means $z_{r-1} = 0$, and this implies
$$z_r = 0 \text{ for all } r \neq 0 \text{ , i.e. } z = z_0 T^0 = z_0 \in K \text{ .}$$
Finally, if σ has the finite order n, we proceed as follows: take in (5) $S := 1$, $a_1 \in L$ such that $\sigma^{r-1}(a_1) \neq a_1$ for all $r \not\equiv 1 \pmod{n}$, and $a_j := 0$ for $j \geq 2$. Just like in the previous case it follows that $z_{r-1} \sigma^{r-1}(a_1) = a_1 \sigma(z_{r-1}) = a_1 z_{r-1}$ for all these r ; again by our choice of a_1 this amounts to $z_{r-1} = 0$, hence
$$z_r = 0 \text{ for all } r \not\equiv 0 \pmod{n} \text{ , i.e. } z \in K((T^n)) \text{ .}$$
It remains to calculate the dimension $|D:Z(D)|$ in the latter case: first we observe $|L:K| = n$ (see Galois Theory, in particular Artin's Lemma (cf. also §6.) for L/K is obviously cyclic with generating automorphism σ . Now choose a basis $\{1, t, t^2, \ldots, t^{n-1}\}$ of L as a K-space, then our considerations show immediately that $\{ t^i T^j \mid 0 \leq i, j < n \}$ is a basis of D as a $K((T^n))$-space, hence $|D:Z(D)| = n^2$. □

Let us note that so far we have only seen skew fields D such that the dimension $|D:Z(D)|$ is either infinite or a square. Later we shall learn that nothing else is possible (cf. §5.).

Exercise 1. Call $a1$ (resp. $bi + cj + dk$) the scalar (resp. pure) component of a quaternion $a1 + bi + cj + dk \in H$ ($a,b,c,d \in R$) and identify the scalar (resp. pure) quaternions - i.e. those with vanishing pure (resp. scalar) component - with the elements in R (resp. R^3). Now show that the scalar (resp. pure) component of the product of two pure quaternions equals the negative scalar product (resp. the vector product) of these two quaternions (viewed as vectors in R^3).

Exercise 2. Study the first two chapters of P.M. Cohn [1977].

§ 2 . RINGS OF MATRICES OVER SKEW FIELDS

Let R be a ring with $1 \neq 0$. We shall henceforth deal with right(left) R-modules $M \neq \{0\}$.

Definition 1. *A right(left) R-module M is called "simple" (or "irreducible") if M contains no proper right(left) R-submodules; M is called "right(left) Noetherian[Artinian]" if every increasing[decreasing] sequence of right(left) R-submodules of M is necessarily finite.*

Definition 2. *If in Definition 1 we are in the special case $M = R$, then we say "right(left) ideal of R" rather than right(left) R-submodule of R; also we say "minimal" right(left) ideal rather than simple right(left) ideal.*

Now let D be a skew field and consider the full matrix ring $M_n(D)$; it is an n^2-dimensional right(left) vector space over D with basis

$$e_{ij} := \begin{pmatrix} & \overset{j\text{-th column}}{\downarrow} & \\ & 1 & \end{pmatrix} \leftarrow i\text{-th row} \qquad (1 \leq i,j \leq n)$$

containing one 1 and n^2-1 0's.

The above basis elements multiply according to

$$e_{ij}e_{rs} = \begin{cases} 0 & j \neq r \\ e_{is} & j = r \end{cases} \text{ in case } .$$

Call $1 := \sum_{i=1}^{n} e_{ii}$ the unit matrix. Then we may consider the *elementary matrices*

$$E_{ij}(t) := 1 + te_{ij} = \begin{pmatrix} 1 & & \overset{j\text{-th column}}{\downarrow} \\ & \ddots & \\ t & & 1 \end{pmatrix} \leftarrow i\text{-th row}$$

$(t \in D ; i \neq j ; 1 \leq i,j \leq n)$.

It is well-known from Linear Algebra that these elementary matrices have the following properties (most of which are fairly obvious):

(1) $E_{ij}(t)E_{ij}(t') = E_{ij}(t+t')$, $E_{ij}(0) = 1$, $E_{ij}(t)^{-1} = E_{ij}(-t)$,

(2) $E_{ij}(t)E_{rs}(t') = E_{rs}(t')E_{ij}(t)$ ($j \neq r \neq s \neq i \neq j$)

and

(3) $E_{ij}(tt') = E_{ir}(t)^{-1}E_{rj}(t')^{-1}E_{ir}(t)E_{rj}(t')$ ($r \neq i \neq j \neq r$) .

Furthermore we introduce matrices

$$D_i(u) := 1 + (u-1)e_{ii} = \begin{pmatrix} 1 & & & 0 \\ & \ddots & & \\ & & u & \\ 0 & & & \ddots \\ & & & & 1 \end{pmatrix} \begin{matrix} \leftarrow \text{i-th column} \\ \\ \leftarrow \text{i-th row} \end{matrix}$$

($u \in D^*$; $1 \leq i \leq n$)

with the properties

(4) $D_i(u)D_i(u') = D_i(uu')$, $D_i(1) = 1$, $D_i(u)^{-1} = D_i(u^{-1})$

and

(5) $D_i(u)D_j(u') = D_j(u')D_i(u)$ ($i \neq j$) .

Finally we assign to every permutation $\pi \in S_n$ of n ciphers a matrix

$$P(\pi) := \left(\delta_{i,\pi(j)} \right) \begin{matrix} \leftarrow \text{i-th row} \\ \\ \end{matrix}$$
$$\uparrow \text{j-th column}$$

where $\delta_{i,\pi(j)} = \begin{matrix} 1 & \text{if} & i = \pi(j) \\ 0 & & i \neq \pi(j) \end{matrix}$.

Obviously in each row and each column of such a *permutation matrix* there is exactly one 1 and n-1 0's . Moreover one checks easily

(6) $P(\pi)P(\pi') = P(\pi\pi')$, $P(id) = 1$, $P(\pi)^{-1} = P(\pi^{-1}) = P(\pi)^t$.

In this context the following is well-known from Linear Algebra:

Lemma 1 . *Given $A \in M_n(D)$ then the "elementary row(column) operations" are as follows:*

transforming from A to	amounts to
$E_{ij}(t)A$ ($AE_{ji}(t)$)	*adding the left(right) t-multiple of the j-th row(column) to the i-th row(column) ;*
$D_i(u)A$ ($AD_i(u)$)	*multiplying the i-th row(column) from the left(right) by u ;*
$P(\pi)A$ ($AP(\pi)^{-1}$)	*moving the i-th row(column) into the position of the $\pi(i)$-th row(column) .*

□

Example 1. Transforming from A to $\begin{pmatrix} 0 & 1 \\ 1 & 0 \end{pmatrix} A \begin{pmatrix} 0 & 1 \\ 1 & 0 \end{pmatrix}$ amounts to rotating the matrix A by 180 degrees (Note: this is different from going over to the transpose A^t).

Example 2.
If we have $A = \left(a_{ij} \right)$ ← i-th row, with j-th column indicated, then we get for every $\pi \in S_n$

$$P(\pi)^{-1} A P(\pi) = \left(a_{\pi(i),\pi(j)} \right) \leftarrow \text{i-th row}$$

↑ j-th column

Lemma 2. *Let* $A = M_n(D)$ *be the full matrix ring over a skew field* D; *then the set of matrices in* A *which commute with all elementary matrices* $E_{ij}(t)$ *is exactly the set of matrices of the form* $z1$ *where* $z \in Z(D)$; *in particular: the mapping* $d \mapsto d1$ *from* D *into* A *induces an isomorphism* $Z(D) \simeq Z(A)$.

Proof. Obviously it suffices to prove the first statement. Call $K := Z(D)$ and note that K is a commutative field (cf. Lemma 1 in §1.; the reader should notice that we do not use this fact in the course of this proof). Clearly we get $z1 \in Z(A)$ for all $z \in K$. Conversely assume that the matrix $A = (a_{rs})$ commutes with all matrices $E_{ij}(t)$; then

$$B := (b_{rs}) := AE_{ij}(t) = E_{ij}(t)A .$$

Equating the main diagonal of B gives

$$a_{jj} = b_{jj} = a_{jj} + a_{ji}t \quad \text{for all } t \in D \text{, hence}$$

$$a_{ji} = 0 \quad \text{for all } i \neq j \text{, i.e. } A \text{ is a diagonal matrix.}$$

On the other hand, equating the (i,j)-th position of B gives the relation

$$a_{ij} + ta_{ii} = b_{ij} = a_{jj}t + a_{ij} \quad \text{for all } t \in D \text{, hence}$$

$$z := a_{ii} = a_{jj} \in D \quad \text{for } i \neq j \text{ (take } t = 1 \text{ and note } a_{ij} =$$
$$= 0 \text{ according to the above) and therefore } tz = zt \text{ for all}$$
$$t \in D \text{, i.e. } z \in Z(D) = K .$$

All this implies $A = z1$. □

So far we have not made use of the fact that D is a skew field; in fact we could replace it *mutatis mutandis* by any ring R with $1 \neq 0$. For the rest of this paragraph, however, we have to use the field property of D.

Lemma 3. *Let D be a skew field, then $A = M_n(D)$ is a right(left) Noetherian[Artinian] A-module.*

Proof. Indeed, any right(left) ideal of A is a right(left) vector space over D; hence our assertion is clear for dimensional reasons. □

Definition 3. *Let A be a ring with $1 \neq 0$. We call A a "simple ring" if A is a right Artinian A-module which has no proper two-sided ideals.*

Theorem 1. *Let D be a skew field, then $A = M_n(D)$ is a simple ring.*

Proof. By virtue of Lemma 3 we must show: "if $a \neq \{0\}$ is a two-sided ideal of A, then $1 \in a$". Indeed, let $0 \neq A = (a_{ij}) \in a$; it follows that we have $a_{ij} \neq 0$ for at least one position (i,j), hence

$$e_{rr} = a_{ij}^{-1} e_{ri} A e_{jr} \in a \quad \text{for all } r \ (1 \leq r \leq n),$$

hence

$$1 = \sum_{r=1}^{n} e_{rr} \in a. \quad \square$$

We close this paragraph with a few remarks of general importance but of no importance for us in the course of these lectures (cf. also the end of §3.).

Definition 4. *Let A be a ring with $1 \neq 0$. We call A a "semisimple ring" if the intersection of all maximal right ideals in A equals $\{0\}$.*

Theorem 2. *Let D be a skew field, then $A = M_n(D)$ is a semisimple ring.*

For the proof it clearly suffices to prove the following

Lemma 4. *Let D be a skew field and consider the ring $A = M_n(D)$. Then*

$$\mathcal{r}_i := \left\{ \begin{pmatrix} 0 & \overset{*}{\underset{*}{-}} & 0 \end{pmatrix} \leftarrow \text{i-th row} \right\} \quad (1 \leq i \leq n) \quad \text{is a maximal right ideal}$$

of A.

Proof. Let $a \supset \mathcal{r}_i$, $a \neq \mathcal{r}_i$ be a bigger right ideal; then there must be a position (i,j) such that $a_{ij} \neq 0$ with some matrix $(a_{rs}) \in a$, i.e.

$$a \ni \begin{pmatrix} * & & * \\ & a_{ij} & \\ * & & * \end{pmatrix} = \begin{pmatrix} 0 & & 0 \\ \text{---} & a_{ij} & \text{---} \\ 0 & & 0 \end{pmatrix} + \begin{pmatrix} * & & * \\ 0 & \text{-----} & 0 \\ * & & * \end{pmatrix} \leftarrow \text{i-th row}$$

$$\underbrace{}_{\text{j-th column}}$$

$$=: \quad A \quad + \quad R \quad \text{where } A \in a.$$

Now denote $P(i,j)$ the permutation matrix belonging to the transposition $(i\ j)$; then
$$e_{ii} = Aa_{ij}^{-1}e_{jj}P(i,j) \in a$$
and hence
$$1 = \sum_{r=1}^{n} e_{rr} \in a \quad \text{since} \quad e_{rr} \in \hbar_i \subset a \quad \text{for} \quad r \neq i \quad \text{anyway.} \quad \square\square$$

Exercise 1. Let H be as in §1. and consider the matrices
$$\begin{pmatrix} 1 & i \\ j & k \end{pmatrix}, \begin{pmatrix} 1 & j \\ i & k \end{pmatrix} \in M_2(H) .$$
Show that the first matrix is invertible and that the second matrix (which is the transpose of the first one) is not invertible.

Exercise 2. Prove (e.g. with the aid of Lemma 1)

(7) $\quad \begin{pmatrix} 1 & & 0 \\ t_{21} & \ddots & \\ \vdots & * & \\ t_{n1} & & 1 \end{pmatrix} = \prod_{j=1}^{n-1} \prod_{i=j+1}^{n} E_{ij}(t_{ij})$

and

(8) $\quad \begin{pmatrix} 1 & t_{12} & -t_{1n} \\ & \ddots & * \\ 0 & & 1 \end{pmatrix} = \prod_{i=n-1}^{1} \prod_{j=n}^{i+1} E_{ij}(t_{ij})$,

where $t_{ij} \in R$, R a ring with $1 \neq 0$.

§ 3 . SIMPLE RINGS AND WEDDERBURN'S MAIN THEOREM

Our goal is to prove a sort of converse to Theorem 1 in §2.

Theorem 1 . *Let A be a ring without proper two-sided ideals and let π be a minimal right ideal of A ; if $M \neq \{0\}$ is a right Artinian R-module then $M \simeq \bigoplus_{i=1}^{n} \pi$ as a right A-module (for some suitable n).*

Corollary 1 . *All minimal right ideals of a ring without proper two-sided ideals are isomorphic.*

Corollary 2 . *Every right ideal of a **simple** ring is a direct sum of minimal right ideals.*

Proof. The two corollaries follow easily from the theorem (in Corollary 1 take $M :=$ "a minimal right ideal" ; in Corollary 2 take $M :=$ "a right ideal"). Now let us prove the theorem: consider

$$\{0\} \neq \mathfrak{a} := \sum_{a \in A} a\pi \subseteq A \ ;$$

$\mathfrak{a}$ is a two-sided ideal of A and therefore we have (by assumption) $\mathfrak{a} = A$. It follows

$$M = \sum_{m \in M} mA = \sum_{m \in M} m\mathfrak{a} = \sum_{m \in M} \sum_{a \in A} ma\pi = \sum_{i=1}^{n} m_i \pi$$

with some *minimal* n such that the last equation holds (the existence of such an n is a consequence of the Artinian property of M). This means in particular, that the right A-module homomorphism f below is *surjective*:

$$f: \bigoplus_{i=1}^{n} \pi \longrightarrow M \ , \ (r_1,\ldots,r_n) \mapsto \sum_{i=1}^{n} m_i r_i \ .$$

But f is also *injective*, for if $\sum_{i=1}^{n} m_i r_i = 0$ and (say) $0 \neq r_1 \in \pi$, hence $r_1 A = \pi$ (since π is minimal), and therefore $m_1 \pi = m_1 r_1 A \subseteq \subseteq \sum_{i=1}^{n} m_i \pi$, contradicting the minimal choice of n . □

Let R be a ring with $1 \neq 0$; then we call an element $e \in R$ an *idempotent* if $e^2 = e$ (e.g. $e = 1$ or $e = 0$). Then obviously eR (resp. Re) is a right (resp. left) ideal of R and $eRe \subseteq eR \cap Re$ is a ring contained in R with unit element e (not a subring unless $e = 1$).

Lemma 1. *Let R be a ring, $e \in R$ an idempotent and L_x the left multiplication with x. Then we have an isomorphism of rings*
$$L: eRe \simeq \mathrm{End}_R(eR) \; , \; eae \mapsto L_{eae} \; .$$

Proof. L is clearly a homomorphism of rings (this is shown by a direct verification). It is *injective*, for if $eaeer = 0$ for all $r \in R$ then (take $r = e$) $eae = eae^3 = 0$. But L is also *surjective*, for let f be an arbitrary right R-module endomorphism of eR, then $f(e) = ea$ for some suitable $a \in R$, and we get (for all $r \in R$)
$$f(er) = f(eer) = f(e)er = eaer = eaeer = L_{eae}(er) \; . \; \square$$

Schur's Lemma. *Let M be a simple right R-module (e.g. a minimal right ideal of R). Then $\mathrm{End}_R(M)$ is a skew field.*

Proof. Let $0 \neq f \in \mathrm{End}_R(M)$ (note that such an f exists because of $M \neq \{0\}$ by virtue of our definition of a simple module) ; both $\mathrm{Ker}\, f$ and $\mathrm{Im}\, f$ are right R-submodules of M, hence $\mathrm{Ker}\, f = \{0\}$ and $\mathrm{Im}\, f = M$, i.e. f is an automorphism of right R-modules, hence f^{-1} exists. $\square$

We need one more technical result:

Lemma 2. *Let M be a right R-module, then*
$$\mathrm{End}_R(\bigoplus_{i=1}^n M) \simeq M_n(\mathrm{End}_R(M)) \; .$$

Proof. Set $S := \bigoplus_{i=1}^n M$, $s := (m_1,..,m_n)$ and define projections $\pi_i: S \to M$, $s \mapsto m_i$ and injections $\iota_i: M \to S$, $m \mapsto (0,..,m,..,0)$ (where the m stands in the i-th position). A simple calculation shows that the above projections/injections satisfy the following equations:
$$\pi_i \iota_j = \begin{cases} \mathrm{id}_M & i = j \\ 0 & i \neq j \end{cases} \quad \text{and} \quad \sum_{r=1}^n \iota_r \pi_r = \mathrm{id}_S \; .$$

Now consider an arbitrary $f \in \mathrm{End}_R(S)$; we find $\pi_i f \iota_j \in \mathrm{End}_R(M)$, and hence we have an additive map
$$\alpha: \mathrm{End}_R(S) \to M_n(\mathrm{End}_R(M)) \; , \; f \mapsto (\pi_i f \iota_j) \; .$$

We claim that α is in fact a homomorphism of rings; indeed, we have

$$\alpha(fg) = (\pi_i fg\iota_j) = (\pi_i f(\sum_{r=1}^{n} \iota_r \pi_r) g\iota_j) =$$

$$(\sum_{r=1}^{n}(\pi_i f\iota_r)(\pi_r g\iota_j)) = \alpha(f)\alpha(g) \quad \text{and} \quad (\text{id}_S) = \text{id}_{M^l} .$$

Conversely, let $f_{ij} \in \text{End}_R(M)$ be given and consider the additive map

$$\beta: M_n(\text{End}_R(M)) \to \text{End}_R(S) , \quad (f_{ij}) \mapsto \sum_{i=1}^{n}\sum_{j=1}^{n} \iota_i f_{ij} \pi_j .$$

Obviously $\beta(\text{id}_{M^l}) = \text{id}_S$, and we have

$$\beta((f_{ir})(g_{rj})) = \beta((\sum_{r=1}^{n} f_{ir}g_{rj})) =$$

$$= \sum_{i=1}^{n}\sum_{j=1}^{n}\sum_{r=1}^{n} \iota_i f_{ir} g_{rj} \pi_j =$$

$$= (\sum_{i=1}^{n}\sum_{r=1}^{n} \iota_i f_{ir} \pi_r)(\sum_{j=1}^{n}\sum_{s=1}^{n} \iota_s g_{sj} \pi_j) = \beta((f_{ir})(g_{sj})) ,$$

i.e. β is also a homomorphism of rings. But now we are done, because it is clear from our definitions that we have $\alpha\beta$ = identity and $\beta\alpha$ = identity. □

Now we are ready for the statement and proof of the first basic theorem in these lectures.

Wedderburn's Main Theorem. *A is a simple ring if and only if one has $A \simeq M_n(D)$ with a skew field D (unique up to isomorphism) and a suitable (unique) n. More precisely: if A is a simple ring and $0 \neq e \in A$ an idempotent (e.g. e = 1), then:*

(i) *all minimal right ideals of A are isomorphic ;*

(ii) *$eAe \simeq M_m(D)$ where $D := \text{End}_A(\hbar)$, $\hbar$ being any minimal right ideal of A ;*

(iii) *D according to (ii) is a skew field such that $Z(D) \simeq Z(A)$, in particular $Z(A)$ is a commutative field ;*

(iv) *$M_n(D) \simeq M_m(E)$ with skew fields D and E implies m = n and $D \simeq E$.*

Proof. We know already from Theorem 1 in §2. that $M_n(D)$ is a simple ring provided D is a skew field. To prove the converse it clearly suffices to show (i),..,(iv). (i) is known from Corollary 1; as for (ii), $e \neq 0$ implies $eA \neq \{0\}$, hence (use Lemma 1, Corollary 2 and Lemma 2)

$$eAe \simeq \text{End}_A(eA) \simeq \text{End}_A(\bigoplus_{i=1}^{m} \hbar) \simeq M_m(D) \quad \text{by definition of D.}$$

(iii) follows then from Schur's Lemma together with Lemma 2 (§2.) and

Lemma 1 (§1.). Let us finally prove *(iv)*: consider the idempotent $0 \neq e_{11} \in M_n(D)$; a straightforward calculation shows $e_{11}M_n(D)e_{11} \simeq D$, i.e. $e_{11}M_n(D)$ is a minimal right ideal of $M_n(D)$ (for otherwise the ring $e_{11}M_n(D)e_{11}$ would be at least a 2×2 matrix ring by virtue of Theorem 1 and Lemmas 1/2). The same is true for the right ideal $e_{11}M_m(E)$ in the other matrix ring $M_m(E)$. Because of $M_n(D) \simeq M_m(E)$ we obtain immediately (use Corollary 1 together with Lemma 1)

$$D \simeq \text{End}_{M_n(D)}(e_{11}M_n(D)) \simeq \text{End}_{M_m(E)}(e_{11}M_m(E)) \simeq E$$

and hence $m = n$ for dimensional reasons. □

Wedderburn's Main Theorem has some obvious consequences; we present a short list below:

Theorem 2. *If A is a simple ring then it is automatically left Artinian as well as right and left Noetherian as an A-module.*

This is now clear from Lemma 3 in §2. □

Theorem 3. *Let A be a simple ring; then the following statements are equivalent:*

(a) A *is a skew field ;*
(b) A *has no zero-divisors $\neq 0$;*
(c) A *has no idempotents $\neq 0,1$;*
(d) A *has no nilpotent elements $\neq 0$.*

Proof. Obvious . □

Finally we may state (cf. Theorem 2 in §2.)

Theorem 4. *Any simple ring is in particular a semisimple ring.* □

Of course one may prove the last theorem directly - i.e. without making use of Wedderburn's Main Theorem - since the intersection of all maximal right ideals is itself a two-sided ideal.

Exercise 1. Let A be a simple ring and M a (say left) A-module. Show that M is a projective A-module (i.e. a direct summand of a free module).

Exercise 2. Let A_i ($i = 1,..,n$) be simple rings; consider the ring $\bigoplus_{i=1}^{n} A_i =: A$ (with componentwise addition and multiplication) and show that A is a semisimple ring.

Exercise 3 . Prove: if A is a ring without proper two-sided ideals, and A has a minimal right ideal $\hbar$ then A is a simple ring.

§ 4. A SHORT CUT TO TENSOR PRODUCTS

Throughout this paragraph let R be a ring with 1 (not necessarily $1 \neq 0$), $P, Q, \ldots$ right R-modules, $A, B, \ldots$ left R-modules and $X, Y, \ldots$ Z-modules (i.e. abelian groups). A Z-bilinear map $f: P \times A \to X$ is then called R-*balanced* if $f(pr, a) = f(p, ra)$ for all $p \in P$, $r \in R$, $a \in A$.

Definition 1. *Let P, A be as above; a pair (T, t) with a Z-module $T = T(P, A)$ and an R-balanced Z-bilinear map $t: P \times A \to T$ is called a "tensor product (over R)" if the following holds: given any R-balanced Z-bilinear $f: P \times A \to X$ then there exists exactly one Z-homomorphism $L_f: T \to X$ such that $f = L_f t$, i.e. the diagram shown commutes.*

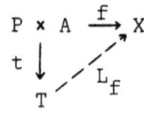

Lemma 1. *Suppose two tensor products (T, t) and (T', t') exist; then $L_{t'}: T \to T'$ is an isomorphism with inverse mapping $L_t: T' \to T$, i.e. $(T', t') = (L_t(T), L_{t'} t)$.*

Proof. Using Definition 1 three times, we get $t = \mathrm{id}_T t$, $t = L_t t'$, $t' = L_{t'} t$ and therefore $t = L_t L_{t'}$, hence $\mathrm{id}_T = L_t L_{t'}$ for uniqueness reasons; in the same way one gets $\mathrm{id}_{T'} = L_{t'} L_t$. □

Lemma 2. *Suppose the tensor product (T, t) exists; then $T = \langle t(P \times A) \rangle$, i.e. T is generated (as a Z-module) by the images under t.*

Proof. Denote $i: \langle t(P \times A) \rangle \to T$ the inclusion. Using Definition 1 with $X = \langle t(P \times A) \rangle$ and $f = t$ gives $t = L_t t$ and therefore $t = it = iL_t t$. On the other hand we have $t = \mathrm{id}_T t$, hence $iL_t = \mathrm{id}_T$ for uniqueness reasons. The latter implies the surjectivity of the inclusion i. □

Following Lemmas 1/2 it is reasonable (and customary) to write $P \otimes_R A$

rather than $T = T(P,A)$ as well as $p \otimes a$ rather than $t(p,a)$, hence $pr \otimes a = p \otimes ra$ for all $p \in P$, $r \in R$, $a \in A$. The commutativity of the diagram in Definition 1 means then $f(p,a) = L_f(p \otimes a)$ and Lemma 2 amounts to the following: given $x \in P \otimes_R A$ then $x = \sum p_i \otimes a_i$ (finite sum). In this context we must be very cautious because $p \otimes a = p \otimes a'$ generally does *not* imply $a = a'$ (cf. Example 1 further below) !

Hitherto it is not clear whether or not $P \otimes_R A$ exists for any given right R-module P and left R-module A. This question will be discussed now.

Theorem 1. *Let* $\pi: P \to Q$ *(* $\alpha: A \to B$ *) be a right(left) R-module homomorphism and assume the existence of both* $P \otimes_R A$ *and* $Q \otimes_R B$. *Then there exists exactly one Z-homomorphism* $\pi \otimes \alpha: P \otimes_R A \to Q \otimes_R B$ *such that* $\pi \otimes \alpha(p \otimes a) = \pi(p) \otimes \alpha(a)$.

Proof: Define the map $f: P \times A \to Q \otimes_R B$, $(p,a) \mapsto \pi(p) \otimes \alpha(a)$. An easy inspection shows that f is R-balanced and Z-bilinear. Now $\pi \otimes \alpha := L_f$ is the required map. □

Theorem 2. $P \otimes_R R$ *exists, more precisely: there is exactly one Z-homomorphism* $\phi: P \otimes_R R \to P$ *such that* $\phi(p \otimes r) = pr$ *for all* $p \in P$, $r \in R$ *and this is an isomorphism with inverse map* $P \to P \otimes_R R$, $p \mapsto p \otimes 1$.

Proof. The map $t: P \times R \to P$, $(p,r) \mapsto pr$ is R-balanced and Z-bilinear. Given now any Z-module X and any R-balanced and Z-bilinear map $f: P \times R \to X$, define a map $L_f: P \to X$, $p \mapsto f(p,1)$ which is clearly a Z-homomorphism such that $f(p,r) = f(pr,1) = L_f(pr) = L_f t(p,r)$. Moreover, because of $t(P \times R) = P$, L_f is unique with this property, hence the pair (P,t) fulfills the requirements for a tensor product according to Definition 1, and we may take $\phi = L_f$. The rest is clear. □

Theorem 3. *Let* $(A_i)_{i \in I}$ *(* $I \neq \emptyset$, *a set of indices) be a family of left R-modules and suppose that* $P \otimes_R A_i$ *exists for every* $i \in I$. *Then* $P \otimes_R (\bigoplus_{i \in I} A_i)$ *exists and there is exactly one isomorphism of Z-modules*

$$P \otimes_R (\bigoplus_{i \in I} A_i) \simeq \bigoplus_{i \in I} (P \otimes_R A_i) \quad \text{such that} \quad p \otimes \sum_i a_i \mapsto \sum_i p \otimes a_i.$$

Proof. Define $t: P \times (\bigoplus_{i \in I} A_i) \to \bigoplus_{i \in I} (P \otimes_R A_i)$, $(p, \sum_i a_i) \mapsto \sum_i p \otimes a_i$; t is R-balanced and Z-bilinear. Given now any Z-module X and any R-balanced and Z-bilinear map $f: P \times (\bigoplus_{i \in I} A_i) \to X$, then - if ι_j denotes the j-th canonical injection of A_j into the direct sum - we have for all $j \in I$ the R-balanced and $\mathbb{Z}$-bilinear maps $f_j: P \times A_j \to X$, $(p, a_j) \mapsto f(p, \iota_j(a_j))$. Now define $L_f: \bigoplus_{i \in I}(P \otimes_R A_i) \to X$, $\sum_i c_i \mapsto \sum_i L_{f_i}(c_i)$. By construction we have $f = L_f t$, and L_f is unique with this property because of $\langle t(P \times (\bigoplus_{i \in I} A_i)) \rangle = \bigoplus_{i \in I}(P \otimes_R A_i)$ (use Lemma 2 for every $i \in I$). Hence the pair $(\bigoplus_{i \in I}(P \otimes_R A_i), t)$ fulfills the requirements for a tensor product according to Definition 1 and Lemma 1 furnishes the isomorphism described in the theorem. □

As an immediate consequence of Theorems 2/3 we obtain

Corollary 1. *If F is a free left R-module then $P \otimes_R F$ exists.* □

Theorem 4. *Let $A \xrightarrow{\alpha} B \xrightarrow{\beta} C \longrightarrow 0$ be an exact sequence of left R-modules and assume the existence of $P \otimes_R A$ and $P \otimes_R B$. Then $P \otimes_R C$ exists, and we have the exact sequence of $\mathbb{Z}$-modules*

$$P \otimes_R A \xrightarrow{id \otimes \alpha} P \otimes_R B \xrightarrow{id \otimes \beta} P \otimes_R C \longrightarrow 0,$$

i.e. "$\otimes$" commutes with "Coker".

Corollary 2. $P \otimes_R A$ *always exists.*

Proof. The corollary is clear, since for given A there are always two free left R-modules F_0, F_1 such that the sequence $F_1 \to F_0 \to A \to 0$ is exact. Hence we get our assertion from Theorem 4 together with Corollary 2. Let us now give a proof of the theorem: set

$$T := \text{Coker } id_P \otimes \alpha = P \otimes_R B \,/\, id_P \otimes \alpha(P \otimes_R A) \,;$$

denote $\bar{z}$ the elements in T, $z \in P \otimes_R B$ being a representative of such a class. Define

$$t: P \times C \to T, \quad (p,c) \mapsto \overline{p \otimes b} \text{ where } b \in B \text{ is such that}$$
$$\beta(b) = c.$$

t is well-defined, for if $\beta(b) = c = \beta(b')$ then $b' - b = \alpha(a)$ for some $a \in A$, hence $p \otimes b' = p \otimes b + id_P \otimes \alpha(p \otimes a)$. Moreover our construction yields $T = \langle t(P \times C) \rangle$ (apply Lemma 2 to $P \otimes_R B$). Now let X be any $\mathbb{Z}$-module and $f: P \times C \to X$ any R-balanced and $\mathbb{Z}$-bilinear map, then the assignment $(p,b) \mapsto f(p, \beta(b))$ defines an R-balanced and $\mathbb{Z}$-bilinear map $g: P \times B \to X$. Now define

$$L_f: T \to X, \quad \bar{z} \mapsto L_g(z) \quad (\bar{z} \in T, \text{ i.e. } z \in P \otimes_R B).$$

L_f is well-defined, for if $\bar{z} = \overline{z'}$ then $z' - z = id_P \otimes \alpha(y)$ for some

$y \in P \otimes_R A$. Lemma 2 gives $y = \sum p_i \otimes a_i$ (finite sum), hence $L_g(z') =$
$= L_g(z) + L_g(id_P \otimes \alpha(\sum p_i \otimes a_i)) = L_g(z) + L_g(\sum p_i \otimes \alpha(a_i)) = L_g(z) +$
$+ \sum f(p_i, \beta\alpha(a_i)) = L_g(z)$. Summarizing we have

$$L_f t(p,c) = L_f(p \otimes b) = L_g(p \otimes b) = f(p, \beta(b)) = f(p,c) ,$$

i.e. $f = L_f t$ and L_f is unique with this property (use Lemma 2). Hence, thanks to Lemma 1, $T \simeq P \otimes_R C$ and $id_P \otimes \beta(p \otimes b) = p \otimes \beta(b) = t(p, \beta(b)) =$
$= \overline{p \otimes b}$, i.e. the map $z \to \overline{z}$ is the same as $id_P \otimes \beta$. □

Example 1. Call M_b the multiplication by b, and consider the exact sequence

$$0 \longrightarrow Z \xrightarrow{M_b} Z \longrightarrow Z/bZ \longrightarrow 0$$

of Z-modules. Thanks to Theorem 4 the upper horizontal sequence of the diagram below is also exact:

$$\begin{array}{ccccccc}
Z/aZ \otimes_Z Z & \xrightarrow{id \otimes M_b} & Z/aZ \otimes_Z Z & \longrightarrow & Z/aZ \otimes_Z Z/bZ & \longrightarrow & 0 \\
\downarrow & & \downarrow & & \downarrow & & \\
Z/aZ & \xrightarrow{M_b} & Z/aZ & \longrightarrow & Z/(a,b)Z & \longrightarrow & 0
\end{array}$$

Moreover the square to the left is commutative, the vertical maps being the isomorphisms from Theorem 2. Finally the lower horizontal sequence is exact as is well-known from Elementary Number Theory (of course (a,b) denotes the greatest common divisor of a and b). Therefore we have an isomorphism

$$Z/aZ \otimes_Z Z/bZ \simeq Z/(a,b)Z ,$$

in particular $Z/aZ \otimes_Z Z/bZ = \{0\}$ provided a and b are coprime !

In the language of categories and functors (which in principle we shall not assume to be known in these lectures) one may state the results which we have achieved so far as follows:

Theorem 5. *The assignment $A \mapsto P \otimes_R A$ (resp. $P \mapsto P \otimes_R A$) for fixed P (resp. A) defines a covariant and right exact functor from the category of left (resp. right) R-modules into the category of Z-modules.* □

If the above functor is *exact*, i.e. (e.g. in the first case) if for all exact sequences $0 \longrightarrow A \xrightarrow{\alpha} B$ of left R-modules the associated sequence $0 \longrightarrow P \otimes_R A \xrightarrow{id \otimes \alpha} P \otimes_R B$ remains exact, then we call P a *flat* right R-module; flat left R-modules are defined accordingly. By virtue of Theorems 2/3 we get immediately

Theorem 6. *Any free module (e.g. a vector space over a (skew) field) is*

a flat module. □

Definition 2. *Let* M *be a left* R*-module as well as a right* S*-module. If we have* $r(ms) = (rm)s$ *for all* $r \in R, m \in M, s \in S$, *then we call* M *an* $(R\text{-}S)$*-bimodule (in short an* R*-bimodule in case* $R = S$*).*

Lemma 3. *Let* A *be a right* R*-module,* M *an* $(R\text{-}S)$*-bimodule and* P *a left* S*-module. Then* $A \otimes_R M$ *(resp.* $M \otimes_S P$*) may be given the structure of a right* S*-module (resp. left* R*-module) such that* $(a \otimes m)s = a \otimes ms$ *(resp.* $r(m \otimes p) = rm \otimes p$*)* ($a \in A, r \in R, m \in M, s \in S, p \in P$). *In this sense there is exactly one* Z*-isomorphism*

$$(A \otimes_R M) \otimes_S P \simeq A \otimes_R (M \otimes_S P) \text{ such that } (a \otimes m) \otimes p \mapsto a \otimes (m \otimes p).$$

Proof. Consider for every $s \in S$ the right "multiplication" $R_s: M \to M$, $m \mapsto ms$; this gives a ring antihomomorphism $f: S \to \text{End}_Z(A \otimes_R M)$, $s \mapsto \text{id}_A \otimes R_s$. Now define $xs := f(s)(x)$ ($x \in A \otimes_R M$); this endows $A \otimes_R M$ with the structure of a right S-module as claimed. A similar procedure leads to the left R-module structure of $M \otimes_S P$. Now consider, for fixed $p \in P$, the Z-bilinear map $h_p: A \times M \to A \otimes_R (M \otimes_S P)$, $(a,m) \mapsto a \otimes (m \otimes p)$. This map is R-balanced (cf. the first part of this lemma), and therefore it induces a Z-homomorphism $f_p: A \otimes_R M \to A \otimes_R (M \otimes_S P)$ which is uniquely determined by the property $a \otimes m \mapsto a \otimes (m \otimes p)$; therefore we may define a map $f: (A \otimes_R M) \times P \to A \otimes_R (M \otimes_S P)$, $(x,p) \to f_p(x)$ which is easily seen to be S-balanced and Z-bilinear. Consequently we have exactly one Z-homomorphism $L_f: (A \otimes_R M) \otimes_S P \to A \otimes_R (M \otimes_S P)$ such that $(a \otimes m) \otimes p \mapsto a \otimes (m \otimes p)$. In the same way one establishes the inverse map, hence L_f is the isomorphism we were looking for. □

Now let us make further investigations in the special (but highly important) case where R is a *commutative* ring; of course, in this case we do not need to distinguish between right and left R-modules for we have R-bimodules such that $rx = xr$ ($r \in R$) which we shall call (as usual) R-modules.

Theorem 7. *Let* R *be commutative and* P, A R*-modules; then*

(A) $P \otimes_R A$ *may be given the structure of an* R*-module such that* $r(p \otimes a) = rp \otimes a = pr \otimes a = p \otimes ra = p \otimes ar = (p \otimes a)r$ ($r \in R, p \in P, a \in A$), *i.e.* $t: (p,a) \mapsto p \otimes a$ *is* R*-bilinear;*

(B) *if in* Definition 1 *the* Z*-module* X *is even an* R*-module and if* f *is* R*-bilinear, then* L_f *is an* R*-module homomorphism;*

(C)   in the situation of Theorem 1 $\pi \otimes \alpha$ is an R-module homomorphism .

Proof. (A) follows from the first part of Lemma 3. (B) and (C) follow by inspection. □

Lemma 4 . Let R be commutative, then there is exactly one isomorphism of R-modules $A \otimes_R B \simeq B \otimes_R A$ such that $a \otimes b \mapsto b \otimes a$.

Proof. Consider the R-bilinear map $f: A \times B \to B \otimes_R A$, $(a,b) \mapsto b \otimes a$. Then Theorem 7 implies the existence of an R-module homomorphism $L_f: A \otimes_R B \to B \otimes_R A$ which is uniquely determined by the property $a \otimes b \mapsto b \otimes a$. Interchanging the roles of A and B gives the inverse map, hence L_f is the claimed isomorphism. □

Another (rather obvious) consequence of our investigations is

Lemma 5 . Let R be commutative and F_m , F_n free R-modules of rank m and n . Then $F_m \otimes_R F_n$ is a free R-module of rank mn . □

An important feature of the case of a commutative ring R is the possibility to define a tensor product of more than two modules.

Definition 3 . Let R be commutative and $M_1,\ldots,M_n$ ($n \geq 2$) R-modules. Define

$$\bigotimes_{j=1}^{2} M_j := M_1 \otimes_R M_2 \quad \text{and inductively} \quad \bigotimes_{j=1}^{n} M_j := M_1 \otimes_R \left(\bigotimes_{j=2}^{n} M_j\right)$$

for $n > 2$; in case $M_1 = \ldots = M_n =: M$ write $M^{\otimes n}$ in place of $\bigotimes_{j=1}^{n} M_j$ with the additional conventions $M^{\otimes 1} := M$ and $M^{\otimes 0} := R$.

Theorem 8 . Let R be a commutative ring, $M_1,\ldots,M_n$ ($n \geq 2$) R-modules, $T := \bigotimes_{j=1}^{n} M_j$ (according to Definition 3) , and $t: M_1 \times \ldots \times M_n \to T$, $(m_1,\ldots,m_n) \mapsto m_1 \otimes (m_2 \otimes \ldots \otimes m_n) =: m_1 \otimes \ldots \otimes m_n$. Then T is an R-module, t is R-multilinear, and the pair (T,t) has the following property: given any R-multilinear map $f: M_1 \times \ldots \times M_n \to X$ then there exists exactly one R-module homomorphism $L_f: T \to X$ such that $f = L_f t$, i.e. the diagram shown commutes.

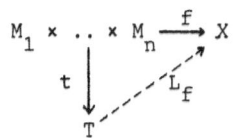

Proof. The first two assertions follow from Theorem 7. As for the universal property of the pair (T,t) we proceed by induction on n, the case $n = 2$ being clear. Therefore let $n > 2$ and fix $m \in M_1$; then the map $f_m: M_2 \times .. \times M_n \longrightarrow X$, $(m_2,...,m_n) \mapsto f(m,m_2,..,m_n)$ is R-multilinear and hence (by the induction hypothesis) there is exactly one R-homomorphism $L_m: \bigotimes_{j=2}^{n} M_j \to X$ such that $L_m(m_2 \otimes .. \otimes m_n) = f(m,m_1,..,m_n)$. Now consider the R-bilinear map $g: M_1 \times (\bigotimes_{j=2}^{n} M_j) \to X$, $(m,y) \mapsto L_m(y)$ which furnishes exactly one R-module homomorphism $L_g: T \to X$ such that $L_g(m \otimes y) =$ $= g(m,y) = L_m(y)$, in particular $L_g(m_1 \otimes (m_2 \otimes .. \otimes m_n)) = L_{m_1}(m_2 \otimes .. \otimes m_n) =$ $= f(m_1,..,m_n)$. Thus we can take $L_f := L_g$. □

Exercise 1. Show that Lemmata 1/2 and Theorem 1 hold *mutatis mutandis* in case of the n-fold tensor product according to Definition 3.

Exercise 2. Let R be a commutative ring, a,b ideals of R and M an R-module. Show $R/a \otimes_R M \simeq M/aM$ and $R/a \otimes_R R/b \simeq R/(a+b)$ (cf. Example 1).

§ 5. TENSOR PRODUCTS AND ALGEBRAS

Let R be a *commutative* ring, A a ring (both with $1 \neq 0$) and $i_A: R \to Z(A) \subseteq A$ a ring homomorphism. As usual we call A an *R-algebra* and write ra instead of $i_A(r)a$ ($r \in R$, $a \in A$). A ring homomorphism $f: A \to B$ is then called an *R-algebra homomorphism* if $i_B = fi_A$, i.e. if the diagram shown commutes.

Theorem 1. *Let A,B be R-algebras; then $A \otimes_R B$ may be given in a unique way the structure of an R-algebra with unit element $1 \otimes 1$ such that $(a \otimes b)(a' \otimes b') = aa' \otimes bb'$ ($a,a' \in A$; $b,b' \in B$).*

Proof. First the R-bilinear map $(a,a') \mapsto aa'$ gives an R-module homomorphism $\alpha: A \otimes_R A \to A$ such that $\alpha(a \otimes a') = aa'$; in the same way we obtain $\beta: B \otimes_R B \to B$. On the other hand we have exactly one R-module isomorphism $h: (A \otimes_R B) \otimes_R (A \otimes_R B) \simeq (A \otimes_R A) \otimes_R (B \otimes_R B)$ such that $(a \otimes b) \otimes (a' \otimes b') \mapsto (a \otimes a') \otimes (b \otimes b')$; this follows by repeated use of Lemmata 3/4 in §4. Now define a multiplication on $A \otimes_R B$ by
$$xy := \alpha \otimes \beta(h(x \otimes y)) \quad (x,y \in A \otimes_R B) ;$$
this multiplication is obviously distributive and inspection shows that it has all the asserted properties. □

The algebras $A \otimes_R B$ have an important feature:

Theorem 2. *Let A,B,C be R-algebras and $f: A \to C$, $g: B \to C$ R-algebra homomorphisms such that $f(a)g(b) = g(b)f(a)$ for all $a \in A$, $b \in B$, then there exists exactly one R-algebra homomorphism $h: A \otimes_R B \to C$ such that $h(a \otimes 1) = f(a)$ and $h(1 \otimes b) = g(b)$ for all $a \in A$, $b \in B$, i.e. the diagram shown is commutative.*

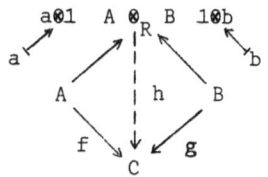

Proof. Consider the R-bilinear map $\phi: A \times B \to C$, $(a,b) \mapsto f(a)g(b)$, then
— thanks to $f(a)g(b) = g(b)f(a)$ — the R-module homomorphism $h := L_\phi$:
$A \otimes_R B \to C$ such that $h(a \otimes b) = f(a)g(b)$ is even an R-algebra homomorphism (check by inspection) and has all the required properties. □

Lemma 1. *Let P be an R-module, A an R-algebra. Then $P \otimes_R A$ may be given in a unique way the structure of a left A-module such that $a'(p \otimes a) = p \otimes a'a$ ($a',a \in A$; $p \in P$). Moreover, if A is commutative and P an R-algebra, then $P \otimes_R A$ is even an A-algebra.*

Proof. Let $L_a: A \to A$ be the left multiplication by $a \in A$. Theorem 7 (C) in §4. then gives a homomorphism of rings
$$h: A \to \mathrm{End}_R(P \otimes_R A), \quad a \mapsto \mathrm{id}_P \otimes L_a.$$
Hence the definition $ax := h(a)(x)$ ($x \in P \otimes_R A$) endows $P \otimes_R A$ with the structure of a left A-module in the required way, uniqueness being implied by Lemma 2 in §4. The rest is clear. □

Lemma 2. *Let P be an R-module, A a commutative R-algebra and B an A-algebra. Then there is exactly one isomorphism of left B-modules $(P \otimes_R A) \otimes_A B \simeq P \otimes_R B$ such that $(p \otimes a) \otimes b \mapsto p \otimes ab$. Moreover, if B is commutative and P an R-algebra, then the above isomorphism is even an isomorphism of B-algebras.*

Proof. Consider, for fixed $b \in B$, the R-bilinear map $f_b: P \times A \to P \otimes_R B$, $(p,a) \mapsto p \otimes ab$; Theorem 7 in §4. then gives an R-module homomorphism $L_b: P \otimes_R A \to P \otimes_R B$ such that $p \otimes a \mapsto p \otimes ab$ and thus an A-bilinear map $f: (P \otimes_R A) \times B \to P \otimes_R B$, $(x,b) \mapsto L_b(x)$. Again this gives an A-module homomorphism $L_f: (P \otimes_R A) \otimes_A B \to P \otimes_R B$ such that $L_f((p \otimes a) \otimes b) = L_b(p \otimes a) = f_b(p,a) = p \otimes ab$, L_f being unique with this property because of Lemma 2 in §4. Inspection shows that L_f is even a homomorphism of left B-modules (in the sense of Lemma 1) and B-algebras (provided P is an R-algebra). It remains to be shown that L_f is bijective; indeed, the R-bilinear map $g: P \times B \to (P \otimes_R A) \otimes_A B$, $(p,b) \mapsto (p \otimes 1) \otimes b$ gives us via Theorem 7 in §4. an R-module homomorphism $L_g: P \otimes_R B \to (P \otimes_R A) \otimes_A B$ which is easily seen to be the inverse of L_f. □

Lemma 3. *Let A,B be R-modules and C a commutative R-algebra. Then there is exactly one isomorphism of C-modules $(A \otimes_R C) \otimes_C (B \otimes_R C) \simeq (A \otimes_R B) \otimes_R C$ such that $(a \otimes c) \otimes (b \otimes c') \mapsto (a \otimes b) \otimes cc'$. Moreover, if A,B are R-algebras, then the above isomorphism is even an isomorphism of C-algebras.*

Proof. By Lemmas 1/2 in connection with Lemma 3 in §4. we have unique R-module isomorphisms $(A \otimes_R C) \otimes_C (B \otimes_R C) \simeq A \otimes_R (B \otimes_R C) \simeq$
$\simeq (A \otimes_R B) \otimes_R C$ such that $(a \otimes c) \otimes (b \otimes c') \mapsto a \otimes (b \otimes cc') \mapsto (a \otimes b) \otimes cc'$. These are even C-module isomorphisms and C-algebra isomorphisms (provided A,B are R-algebras) as is easily seen by inspection. □

Theorem 3. *Let P be an R-module, A an R-algebra, X a left A-module and f: P → X an R-module homomorphism. Then there exists exactly one left A-module homomorphism f_A: P $\otimes_R$ A → X (the so-called "left A-linear extension of f") such that $p \otimes a \mapsto af(p)$. If in addition to the above assumptions P,X are R-algebras and if f is then an R-algebra homomorphism, then f_A is also an R-algebra homomorphism and even an A-algebra homomorphism provided A is commutative and X is an A-algebra.*

Proof. Consider the R-bilinear map g: P × A → X, $(p,a) \mapsto af(p)$ and take $f_A := L_g$. Then (see Lemma 1) $f_A(a'(p \otimes a)) = L_g(p \otimes a'a) = a'af(p) =$
$= a'L_g(p \otimes a) = a'f_A(p \otimes a)$. All the rest is clear after a short calculation.□

An important application of the above theorem is

Lemma 4. *Let A be an R-algebra and $j: M_n(R) \to M_n(A)$ the canonical map induced by $i_A: R \to A$, $r \mapsto r1$. Then the left A-linear extension $j_A: M_n(R) \otimes_R A \to M_n(A)$ is an isomorphism of R-algebras and left A-modules (and A-algebras provided A is commutative).*

Proof. $M_n(R)$ is a free R module with basis $\{e_{ij}\}$ (see §2.), hence - by Theorem 3 in §4 in connection with Lemma 1 - $M_n(R) \otimes_R A$ is a free left A-module with basis $\{e_{ij} \otimes 1\}$, this basis being mapped onto the basis $\{e_{ij}\}$ of the left A-module $M_n(A)$. Hence j_A must be an isomorphism. The rest is clear after a quick inspection. □

Corollary 1. $M_n(R) \otimes_R M_m(R) \simeq M_n(M_m(R)) = M_{nm}(R)$. □

Now we want to use the results hitherto described in order to investigate the tensor product of two simple rings (in the sense of §§2/3.) over a suitable common subring. For this purpose we need some more preparation.

Definition 1. *Let A be a ring; define the "inverse" (or "opposite") ring A^{op} to be the additive group A equipped with the new multiplication "·" such that $a \cdot b := ba$ (the latter being understood in the old sense).*

In this context the following can be observed without any difficulties:

(1) $(A^{op})^{op} = A$;

(2) $A^{op} = A$ if and only if A is commutative ;

(3) $Z(A^{op}) = Z(A)$;

(4) $f: A^{op} \to B$ is a ring homomorphism if and only if $f: A \to B$ is a ring antihomomorphism ;

(5) $a \subseteq A$ is a left(right) ideal if and only $a \subseteq A^{op}$ is a right ideal ;

(6) A is a simple ring if and only A^{op} is a simple ring ;

(7) A is an R-algebra if and only A^{op} is an R-algebra ;

(8) $(A \otimes_R B)^{op} \simeq A^{op} \otimes_R B^{op}$ for R-algebras A, B ;

(9) $a \mapsto a^t$ (:= transpose of a) gives an isomorphism of R-algebras $M_n(A)^{op} \simeq M_n(A^{op})$.

Lemma 5. *Let* A, B *be R-algebras and* $f: A \to B$ *an R-algebra homomorphism. Then there exists exactly one R-algebra homomorphism*

$$\Omega_f: A \otimes_R B^{op} \to End_R(B) \quad such \; that \quad \Omega_f(a \otimes b) = L_{f(a)} R_b \; ,$$

L_x *and* R_x *being the left and right multiplication by* x *on* B .

Proof. Consider the R-algebra homomorphisms $A \to End_R(B)$, $a \mapsto L_{f(a)}$ and $B^{op} \to End_R(B)$, $b \mapsto R_b$ with $L_{f(a)} R_b = R_b L_{f(a)}$ (this amounts to the the associative law in B); now use Theorem 2 and set $\Omega_f := h$. □

Definition 2. *If in* Lemma 5 *we have* $A = B$ *and* $f = id_A$ *then the elements in* $\Omega_{id_A}(A \otimes_R A^{op}) \subseteq End_R(A)$ *are called "analytic R-linear maps of* A " .

Hence $f \in End_R(A)$ is analytic R-linear if and only if

(10) $\quad f(x) = \sum a'_r x a''_r$ for a finite number of suitable $a'_r, a''_r \in A$.

Artin-Whaples Theorem. *Let* A *be a K-algebra without proper two-sided ideals,* $K := Z(A)$ *a (commutative) field,* $a_1, \ldots, a_n \in A$ *linearly independent over* K *and* $b_1, \ldots, b_n \in A$ *arbitrary. Then there exists an analytic K-linear map* f *such that* $f(a_r) = b_r$ ($r = 1, \ldots, n$).

Proof. First we observe that it clearly suffices to prove this theorem in the special case $b_1 = \ldots = b_{n-1} = 0$ and $b_n = 1$. Now let us proceed by induction on n : the case $n = 1$ is trivial, since then $A a_1 A$ is a two-sided ideal $\neq \{0\}$ of A , hence $1 = \sum a'_s a_1 a''_s$ for some $a'_s, a''_s \in A$. Now the case $n > 1$: by induction hypothesis we may select analytic K-linear maps $f_1, \ldots, f_{n-1}$ such that

$$f_i(a_j) = \begin{cases} 1 & i = j \\ 0 & i \neq j \end{cases} \quad \text{if} \quad (1 \leq i,j < n).$$

The values $f_i(a_n)$ are then unknown. Let us first consider the case where $f_m(a_n) \notin Z(A)$ for some index m ($1 \leq m < n$); choose a $\in A$ such that $0 \neq b := f_m(a_n)a - af_m(a_n) \in A$ and set $g(x) := f_m(x)a - af_m(x)$ ($x \in A$); now choose an analytic K-linear map h such that $h(b) = 1$ (cf. the case $n = 1$). Then $f := hg$ is analytic K-linear and by construction we have $f(a_1) = \ldots = f(a_{n-1}) = h(0) = 0$ and $f(a_n) = h(b) = 1$. Now consider the remaining case where $f_i(a_n) \in Z(A) = K$ for all $i = 1,\ldots,n-1$. Define $g(x) := x - \sum_{i=1}^{n-1} f_i(x)a_i$ with $b := g(a_n) \neq 0$, for otherwise the $a_1,\ldots,a_n$ would be linearly dependent over K. Now define h and f as in the previous subcase; again f is analytic K-linear and according to our requirements. □

Corollary 2. *Let* A *be a finite dimensional* K-*algebra without proper two-sided ideals,* $Z(A) =: K$ *a field and* $n := |A:K|$ *, then there is a* K-*algebra isomorphism* $A \otimes_K A^{op} \simeq M_n(K)$.

Proof. The Artin-Whaples Theorem says then that the map Ω_{id_A} from Definition 2 is surjective. After identifying $End_K(A)$ with $M_n(K)$ we see that Ω_{id_A} must also be injective for dimensional reasons. □

An algebra A as in Corollary 2 is of course a simple ring and consequently $Z(A)$ is automatically a field (see §3.). Conversely, every simple ring which is a finite dimensional algebra over its centre is necessarily an algebra of the type discussed in Corollary 2.

Definition 3. *A simple ring* A *with finite* $|A:Z(A)|$ *is called a "central simple* K-*algebra"* ($K := Z(A)$)*. If a central simple* K-*algebra is a skew field then we call it a "*K-*skew field".*

In what follows now we shall make use of the following facts: if K is a commutative *field* and if A,B are K-algebras, then (see Theorem 6 in §4.) we have *injective* K-algebra homomorphisms $A \rightarrow A \otimes_K B \leftarrow B$, $a \mapsto a \otimes 1$ and $1 \otimes b \leftarrow b$, hence we may - and henceforth shall - view A and B as *embedded* in $A \otimes_K B$. In this sense we have:

Theorem 4. *Let* A,B *be* K-*algebras,* $K = Z(A)$ *a field and* A *without proper two-sided ideals. Let* $c \subseteq A \otimes_K B$ *be a two-sided ideal,* $b := $ $:= c \cap B$ *(see remarks above), then* $c = A \otimes_K b$.

Proof. "$\supseteq$" is clear; conversely let $c \ni c = \sum a_j \otimes b_j$ (finite sum) ; we may restrict ourselves to the case where the a_j are linearly independent over K. Now use the Artin-Whaples Theorem and choose f_i such that

$$f_i(a_j) = \begin{cases} 1 & i = j \\ 0 & i \neq j \end{cases}.$$

It follows $f_i \otimes id_B(c) = \sum_j f_i(a_j) \otimes b_j = 1 \otimes b_i \in c \cap B = b$ for all i (cf. the explicit form of f_i as shown in (10)). □

Corollary 3. *Let* A, B *be* K-*algebras over a field* $K := Z(A) \subseteq Z(B)$; *then* $A \otimes_K B$ *has no proper two-sided ideals if and only* A *and* B *have no proper two-sided ideals.*

Proof. If (say) a is a two-sided ideal of A then $a \otimes_K B$ is such an ideal in $A \otimes_K B$. The converse implication is then clear from Theorem 4. □

Before we close this paragraph we need to make a few remarks concerning left Artinian R-modules (of course, a similar remark holds for right modules).

Lemma 6. *Let* R *be a (not necessarily commutative) ring,* M, N *left* R-*modules and* $f: M \to N$ *a left* R-*module homomorphism. Then* M *is left Artinian if and only if* $\text{Ker } f$ *and* $\text{Im } f$ *are left Artinian.*

Corollary 4. *Let* M, N *be left* R-*modules; then* $M \oplus N$ *is Artinian if and only if* M *and* N *are Artinian.*

Proof. The corollary follows from the lemma using the projection onto one factor. Now the proof of the lemma: first we observe that $\text{Ker } f$ and $\text{Im } f$ are clearly Artinian provided M is. As for the converse we note that any chain $M \supseteq M_1 \supseteq M_2 \supseteq \ldots$ of submodules of M gives us two such chains $\text{Im } f \supseteq f(M_1) \supseteq f(M_2) \supseteq \ldots$ and $\text{Ker } f \supseteq M_1' \supseteq M_2' \supseteq \ldots$ where $M_r' := M_r \cap \text{Ker } f$. By assumption we get $f(M_r) = f(M_n)$ and $M_r' = M_n'$ for all $r \geq n$ for some n. Now let $x \in M_n$, hence $f(x) \in f(M_n) \subseteq \subseteq f(M_r)$, i.e. $x = y + z \in M_r + M_n' \subseteq M_r + M_r' = M_r$ for all $r \geq n$. The latter amounts to $M_r = M_n$ for all these r. □

Theorem 5. *Let* A, B *be* K-*algebras,* $K = Z(A) \subseteq Z(B)$ *a field and either* $|A:K|$ *or* $|B:K|$ *finite; then* $A \otimes_K B$ *is a simple ring if and only* A *and* B *are simple rings.*

Proof. Thanks to Corollary 3 we only have to show: $A \otimes_K B$ is a left Artinian ($A \otimes_K B$)-module if and only if A (resp. B) are left Artinian

A-(resp. B-)modules. Here the "only if" is obvious (and holds even *without* the finiteness assumptions) since a strictly decreasing infinite sequence of right ideals a_n of (say) A would give us an infinite such sequence $a_n \otimes_K B$ of $A \otimes_K B$. As for the converse we note $A \otimes_K B \simeq \bigoplus A$ (finite sum) as a left A-module (provided (say) $|B:K|$ is finite); this follows from Theorem 3 in §4. together with Lemma 1. Therefore by Corollary 4 $A \otimes_K B$ is a left Artinian A-module hence even more so a left Artinian $(A \otimes_K B)$-module. □

Now we want to show that the finiteness condition in the previous theorem is not superfluous.

Theorem 6. *Let* A *be a K-algebra without proper two-sided ideals,* $K = Z(A)$ *a field; then* $A \otimes_K A^{op}$ *is a simple ring if and only if* $|A:K|$ *is finite (i.e.* A *is a central simple K-algebra).*

Proof. The "if" has already been proved in Corollary 2 (in connection with Theorem 1 in §2.). Conversely assume $|A:K|$ being infinite and select a sequence $a_1, a_2, a_3, \ldots \in A$ of K-linearly independet elements in A. Now define
$$a_n := \{ f \in \mathrm{End}_K(A) \mid \begin{array}{l} f \text{ analytic K-linear} \\ f(a_1) = \ldots = f(a_n) = 0 \end{array} \} \ .$$

Clearly we have $a_n \supset a_{n+1}$ and the Artin-Whaples Theorem implies even $a_n \neq a_{n+1}$ ($n \in \mathbb{N}$); on the other hand a_n is a left ideal in $\Omega_{\mathrm{id}_A}(A \otimes_K A^{op})$ (cf. Definition 2). But Ω_{id_A} is *injective* thanks to Corollary 3 (the kernel is a two-sided ideal), hence $A \otimes_K A^{op}$ is not left Artinian and therefore cannot be a simple ring. □

We close this paragraph with an important remark.

Theorem 7. *Let* A *be a central simple K-algebra, then* $|A:K|$ *is a square.*

Corollary 5. *Let* D *be a skew field, then* $|D:Z(D)|$ *is either infinite or a square.*

Proof. Let $\overline{K}$ be an algebraically closed field containing K and consider the K-algebra $A \otimes_K \overline{K}$ with finite $|A \otimes_K \overline{K}:\overline{K}| = |A:K|$ (cf. Theorem 3 (§4.) and Lemma 1). By Theorem 5 $A \otimes_K \overline{K}$ is a simple ring, hence Wedderburn's Main Theorem (§3.) gives $A \otimes_K \overline{K} \simeq M_n(D)$ for some skew field D such that $\overline{K} \subseteq Z(D) \subseteq D$ and $|D:\overline{K}|$ is finite. It follows $D = \overline{K}$ and this means $|A:K| = n^2$. □

Exercise 1. Let K be a field and $L = K(\theta)$ an extension field generated by an element θ with minimal polynomial f over K. If F/K is any (not necessarily finite) extension, show that $L \otimes_K F \simeq F[T]/(f)$ and deduce that $L \otimes_K F$ is a semisimple ring provided L/K is a separable extension (cf. Exercise 2 in §3.).

Exercise 2. Let R be a commutative ring and M an R-module. Consider the R-module

$$T_R(M) := \bigoplus_{n=0}^{\infty} M^{\otimes n}$$

and show that it can be endowed with the structure of an R-algebra such that the following holds: there is an R-module homomorphism $t: M \to T_R(M)$ such that to any given R-module homomorphism $f: M \to A$ into an R-algebra A there exist exactly one R-algebra homomorphism $L_f: T_R(M) \to A$ fulfilling $f = L_f t$. $T_R(M)$ is called the "tensor algebra of M over R".

§ 6 . TENSOR PRODUCTS AND GALOIS THEORY

Classical Galois Theory (which we assume to be known to the reader) is founded on two basic results:

Dedekind's Lemma . *Let* G *be a group,* K *a field and* $f_i: G \to K^*$ ($i = 1,..,n$) *distinct group homomorphisms. Then*
$$\sum_{i=1}^{n} f_i(g)x_i = 0 \text{ for all } g \in G \text{ with } x_g \in K$$
implies $x_1 = .. = x_n = 0$.

Artin's Lemma . *Let* G *be a group of automorphisms of a field* L *and* $K := \text{Fix}_L(G)$ *its fixed field, then* $|L:K| = |G|$ *whenever either side is finite.*

For proofs of these two results cf. e.g. vol. 2 of P.M. Cohn [1974/77]. ◻

Now let us draw the usual conclusions therefrom concerning *Galois Cohomology*. We start with a few preparatory remarks.

In what follows let Γ be a group (written multiplicatively, not necessarily finite, though finite in all of our applications) and M a left Γ-*module* - i.e. a Z-module (usually written additively, although written multiplicatively in many of our applications) where $^\sigma m$ is defined for all $\sigma \in \Gamma$, $m \in M$ such that $^1m = m$, $^\sigma(^\tau m) = {}^{\sigma\tau}m$ and $^\sigma(m + m') = {}^\sigma m + {}^\sigma m'$ ($\sigma \tau \in \Gamma$; $m,m' \in M$) - ; if M,N are left Γ-modules and $f: M \to N$ a Z-module homomorphism, then f is called a left Γ-*module homomorphism* if $f(^\sigma m) = {}^\sigma f(m)$ for all $\sigma \in \Gamma$ and $m \in M$. As usual we call $M^\Gamma := \{ m \in M \mid {}^\sigma m = m \text{ for all } \sigma \in \Gamma \}$ the *fixed module* of M ; it is the largest submodule of M on which Γ acts trivially.

If Γ is a *finite* group then
$$N_\Gamma: M \to M \quad , \quad m \mapsto \sum_{\sigma \in \Gamma} {}^\sigma m$$
is a left Γ-module homomorphism called the *norm*.

Definition 1. *Let* Γ *be a finite group and* M *a left* Γ-*module, then* $H^0(\Gamma,M) := M^\Gamma/N_\Gamma(M)$ *is called the "0-th Cohomology Group of* M*".*

Example 1. Let L/K be finite Galois, $\Gamma := \text{Gal}(L/K)$, then L^* and L^+ are Γ-modules by virtue of the definition $^\sigma x := \sigma(x)$ ($\sigma \in \Gamma$, $x \in L$). Thanks to Galois Theory we have $L^{*\Gamma} = K^*$ and $L^{+\Gamma} = K^+$ as well as $N_\Gamma = N_{L/K}$ and $N_\Gamma = \text{Tr}_{L/K}$, hence

$$H^0(\Gamma,L^*) = K^*/N_{L/K}(L^*) \text{ is the } \textit{norm residue class group.}$$

Moreover,

(1) $\qquad H^0(\Gamma,L^+) = K^+/\text{Tr}_{L/K}(L^+) = \{0\}$.

Indeed, thanks to the K-linearity of the trace, it suffices to find an element $x \in L$ such that $\text{Tr}_{L/K}(x) \neq 0$; however, the existence of such an x is obvious from Dedekind's Lemma (take $\{f_i\} = \Gamma$ and $G = L^*$).

Now let M be a left Γ-module, then

$$C^1(\Gamma,M) := \{\, x: \Gamma \to M \mid x(1) = 0 \,\&\, x(\sigma\tau) = x(\sigma) + {}^\sigma x(\tau)\,\}$$

is called the set of 1-*cocycles (of* Γ *with values in* M *)*. C^1 carries the structure of a Z-module (by pointwise definition of the (say) addition) and an easy calculation shows that the 1-*coboundaries*

$$B^1(\Gamma,M) := \{x: \Gamma \to M \mid x(\sigma) = {}^\sigma m - m \text{ for some } m \in M\,\}$$

form a Z-submodule of C^1. Note that in more old-fashioned terminology C^1 resp. B^1 is called the group of *crossed homomorphisms* resp. *principal crossed homomorphisms (from* Γ *into* M *)*.

Definition 2. *Let* Γ *be a group and* M *a left* Γ-*module, then* $H^1(\Gamma,M) := C^1(\Gamma,M)/B^1(\Gamma,M)$ *is called the "1st Cohomology Group of* M *".*

The following results are classical and of great importance later in these lectures.

Noether's Equations. *In the situation of* Example 1 *above we have*

$$H^1(\Gamma,L^*) = \{1\}\,.$$

Proof. Let $x \in C^1(\Gamma,L^*)$ be given and use Dedekind's Lemma (take $\{f_i\} = \Gamma$ and $G = L^*$) in order to find an $a \in L^*$ such that $b := \sum_{\sigma \in \Gamma} x(\sigma)^\sigma a$

$\neq 0$. Consequently we find

$$^\tau b = \sum_{\sigma \in \Gamma} {}^\tau x(\sigma)^{\tau\sigma}a = \sum_{\sigma \in \Gamma} x(\tau)^{-1} x(\tau\sigma)^{\tau\sigma}a = x(\tau)^{-1}b\,,$$

hence - if $m := b^{-1} \in L^*$ -

$$x(\tau) = {}^{\tau}mm^{-1} \quad \text{for all} \quad \tau \in \Gamma, \text{ i.e. } x \in B^1(\Gamma, L^*).\;\square$$

As a consequence of the previous result we get the (much older)

Hilbert's "Satz 90". *Let L/K be cyclic with generating automorphism σ, then any $x \in L$ such that $N_{L/K}(x) = 1$ is necessarily of the form $x = {}^{\sigma}mm^{-1}$.*

Proof. Denote $\Gamma := \mathrm{Gal}(L/K) = \langle \sigma \rangle$ define a 1-cocycle by $x(\sigma^i) := \prod_{j=0}^{i-1} \sigma^j x$ (note that in view of $N_{L/K}(x) = 1$ this is really a reasonable definition of a cocycle in $C^1(\Gamma, L^*)$) and use Noether's Equations: this gives immediately $x = x(\sigma) = {}^{\sigma}mm^{-1}$. $\square$

Both Noether's Equations and Hilbert's "Satz 90" (which happened to be Theorem 90 in D. Hilbert's famous report *Die Theorie der algebraischen Zahlkörper* published in 1897) have an additive counterpart, namely:

Lemma 1. *In the situation of* Example 1 *we have*

$$H^1(\Gamma, L^+) = \{0\}.$$

Corollary 1. *Let L/K be cyclic with generating automorphism σ, then any $x \in L$ such that $\mathrm{Tr}_{L/K}(x) = 0$ is necessarily of the form $x = {}^{\sigma}m - m$.*

Proof. We just copy the proof of the two preceding results *mutatis mutandis*: first let $x \in C^1(\Gamma, L^+)$ be given, choose (using (1)) an element $a \in L$ such that $\mathrm{Tr}_{L/K}(a) = 1$ and define $b := \sum_{\sigma \in \Gamma} x(\sigma)^{\sigma} a$. Again one finds

$$^{\tau}b = \sum_{\sigma \in \Gamma} {}^{\tau}x(\sigma)^{\tau\sigma}a = \sum_{\sigma \in \Gamma}(x(\tau\sigma) - x(\tau))^{\tau\sigma}a = b - x(\tau),$$

hence - if $m := -b \in L$ -

$$x(\tau) = {}^{\tau}m - m \quad \text{for all} \quad \tau \in \Gamma, \text{ i.e. } x \in B^1(\Gamma, L^+).$$

For the proof of the corollary denote $\Gamma := \mathrm{Gal}(L/K) = \langle \sigma \rangle$ and define a 1-cocycle just like in the proof of Hilbert's "Satz 90" (but with "$\sum$" in place of "$\prod$"); by the lemma this cocycle is a coboundary and so we may conclude ${}^{\sigma}x = m - m$. $\square$

Hitherto we have not dealt with tensor products although they appear in the heading of this paragraph.

Definition 3. *Let L/K be finite Galois, $\Gamma := \mathrm{Gal}(L/K)$ and V a (not necessarily finite dimensional) vector space over L which is also a left*

Γ-module; then we shall call V an "L/K-Galois module" if

(2) $\qquad \sigma(xv) = {}^\sigma x\, {}^\sigma v$ for all $\sigma \in \Gamma$, $x \in L$, $v \in V$.

A straightforward calculation shows

Lemma 2. *Let L/K be finite Galois, $\Gamma := \mathrm{Gal}(L/K)$ and V a vector space over K, then $V \otimes_K L$ (with left Γ-module structure via $\mathrm{id}_V \otimes \sigma$) is an L/K-Galois module. Moreover, if two L/K-Galois modules V, W are given, then $V \otimes_L W$ (with left Γ-module structure via $\sigma \otimes \sigma$) is also an L/K-Galois module.* □

Now we are prepared for the statement and proof of

Theorem 1. *Let V be an L/K-Galois module. Then the following holds:*

(I) $\qquad V^\Gamma$ *is a K-space such that* $LV^\Gamma = V$ *and if* $U \subseteq V^\Gamma$ *is any K-subspace such that* $LU = V$, *then* $U = V^\Gamma$; *moreover the L-linear extension (cf. Theorem 3 in §5.) of the embedding* $V^\Gamma \hookrightarrow V$ *is an isomorphism* $V^\Gamma \otimes_K L \simeq V$ *of L-spaces and left Γ-modules.*

(II) $\qquad H^0(\Gamma, V) = \{0\}$, *i.e.* $V^\Gamma = N_\Gamma(V)$.

(III) $\qquad$ *If W is a further L/K-Galois module, then there is exactly one isomorphism of K-spaces* $V^\Gamma \otimes_K W^\Gamma \simeq (V \otimes_L W)^\Gamma$ *such that* $v \otimes_K w \mapsto v \otimes_L w$.

Corollary 2. *In the situation of Theorem 1 we have*
$$\dim_K(V^\Gamma) = \dim_L(V) \text{ whenever either side is finite.}$$

Proof. The corollary follows from (I) by Theorem 3 in §4. As for the proof of the theorem we begin by observing that V^Γ and $N_\Gamma(V)$ are obviously K-spaces. Now suppose $LN_\Gamma(V) \neq V$, then there would exist an L-linear form $f : V \to L$ such that $f \neq 0$ but $f(N_\Gamma(V)) = \{0\}$. Choose $v \in V$ such that $f(v) \neq 0$; it follows
$$0 = f(N_\Gamma(xv)) = f(\sum_{\sigma \in \Gamma} {}^\sigma x\, {}^\sigma v) = \sum_{\sigma \in \Gamma} {}^\sigma x\, f({}^\sigma v) \quad \text{for all} \quad x \in L^*,$$
contradicting Dedekind's Lemma (take $\{f_i\} = \Gamma$ and $G = L^*$). Therefore $V = LN_\Gamma(V) \subseteq LV^\Gamma \subseteq V$, hence $LN_\Gamma(V) = V = LV^\Gamma$, in particular we get (II) as soon as we have finished the proof of (I). Now consider the embedding $i : U \hookrightarrow V$, then (because of our assumption $LU = V$) the L-linear extension $i_L : U \otimes_K L \to V$ (which is such that $u \otimes x \mapsto xu$) is surjective. We want to show its injectivity: for this purpose let $y \in U \otimes_K L$ be given; we write $y = \sum_{j=1}^m v_j \otimes x_j$ with some m and may assume the elements

v_j to be K-linear independent. Now assume $i_L(y) = 0$; we must show $y = 0$. Thanks to $N_\Gamma(L) \subseteq K$ (actually we know even "=" from (1)) we obtain

$$0 = N_\Gamma(xi_L(y)) = \sum_{j=1}^{m} (N_\Gamma(xx_j))v_j \quad \text{for all } x \in L^* \text{ , hence}$$

$$0 = N_\Gamma(xx_j) = \sum_{\sigma \in \Gamma} {}^\sigma x {}^\sigma x_j \quad \text{for all } x \in L^* \quad (1 \leq j \leq m),$$

therefore - using Dedekind's Lemma m times - $x_j = 0$ for all j, i.e. $y = 0$ and consequently $U \otimes_K L \simeq V$. Now let $v \in V^\Gamma$ be given; because of $LU = V$ we may find elements $x_j \in L$ and K-linearly independent elements $u_j \in U$ such that (for some m)

$$\sum_{j=1}^{m} x_j u_j = v = {}^\sigma v = \sum_{j=1}^{m} {}^\sigma x_j u_j \quad \text{for all } \sigma \in \Gamma,$$

hence $x_j \in \text{Fix}_L(\Gamma) = K$ for all j, i.e. $v \in U$ and thus $U = V^\Gamma$. This proves *(I)*. It remains to prove *(III)*: first we note that the map $g: V^\Gamma \times W^\Gamma \to V \otimes_L W$, $(v,w) \mapsto v \otimes_L w$ is obviously K-bilinear; consequently there is exactly one K-homomorphism $L_g: V^\Gamma \otimes_K W^\Gamma \to V \otimes_L W$ such that $v \otimes_K w \mapsto v \otimes_L w$. Now let $\{v_i\}_{i \in I}$ resp. $\{w_j\}_{j \in J}$ be bases of V^Γ resp. W^Γ. Then (cf. §4.) $\{v_i \otimes_K w_j\}$ is a K-basis of the space $V^\Gamma \otimes_K W^\Gamma$, on the other hand $\{v_i \otimes_L w_j\}$ is an L-basis of $V \otimes_L W$ (cf. again §4. together with *(I)* above), hence the elements $v_i \otimes_L w_j = L_g(v_i \otimes_K w_j) \in (V \otimes_L W)^\Gamma$ are even more so K-linearly independent which amounts to the injectivity of L_g. Finally we observe $L_g(V^\Gamma \otimes_K W^\Gamma) \subseteq (V \otimes_L W)^\Gamma$ and $LL_g(V^\Gamma \otimes_K W^\Gamma) = V$ (cf. the argument involving the bases which proved the injectivity), hence $L_g(V^\Gamma \otimes_K W^\Gamma) = (V \otimes_L W)^\Gamma$ by *(I)*. □

Definition 4. *Let* L/K *be finite Galois,* $\Gamma := \text{Gal}(L/K)$ *and* A *an* L*-algebra which is also a left* Γ*-module; then we shall call* A *an* " L/K*-Galois algebra* " *if*

(3) $^\sigma(aa') = {}^\sigma a {}^\sigma a'$ *for all* $\sigma \in \Gamma$; $a, a' \in A$.

Since (3) implies (2) we see that an L/K-Galois algebra is in particular an L/K-Galois module. (3) means that the Galois group Γ can be viewed (by prolongation) as a group of K-algebra automorphisms of the L-algebra given. The following result is an almost obvious supplement to Theorem 1 :

Theorem 2. *If in* Theorem 1 V *and* W *are even* L/K*-Galois algebras, then* V^Γ *and* W^Γ *are* K*-algebras and the isomorphisms of (I) resp. (III) are* L*-algebra resp.* K*-algebra isomorphisms.* □

Exercise 1. Let $\Gamma = \langle \sigma \rangle$ be a finite cyclic group and M a left Γ-module.

Show $\quad H^1(\Gamma,M) \simeq \operatorname{Ker} N_\Gamma / \{ {}^\sigma m - m \mid m \in M \}$.

Exercise 2. Let $0 \longrightarrow M \xrightarrow{f} N \xrightarrow{g} P \longrightarrow 0$ be an exact sequence of Γ-modules and Γ-module homomorphisms. Show that there is a Z-homomorphism $\partial: P^\Gamma \to H^1(\Gamma,M)$ such that there is an exact sequence of Z-modules and Z-homomorphisms (here f^1 resp. g^1 are induced by f resp. g in the evident way)

$0 \longrightarrow M^\Gamma \xrightarrow{f} N^\Gamma \xrightarrow{g} P^\Gamma \xrightarrow{\partial} H^1(\Gamma,M) \xrightarrow{f^1} H^1(\Gamma,N) \xrightarrow{g^1} H^1(\Gamma,P)$.

§ 7. SKOLEM-NOETHER THEOREM AND CENTRALIZER THEOREM

In this paragraph we want to establish two substantial results. We start with

Definition 1. *Let R be a commutative ring, A,B R-algebras, $f: A \to B$ an R-algebra homomorphism and $\Omega_f: A \otimes_R B^{op} \to \mathrm{End}_R(B)$ according to Lemma 5 in §5., then we denote by B_f our R-algebra B viewed as a left $(A \otimes_R B^{op})$-module via $xb := \Omega_f(x)(b)$ ($x \in A \otimes_R B^{op}$, $b \in B$).*

Recall from Lemma 5 in §5. that the above definition gives in particular $(a' \otimes b')b = f(a')bb'$ ($a' \in A$; $b,b' \in B$). The following is crucial:

Theorem 1. *Let A,B be R-algebras and $f,g: A \to B$ R-algebra homomorphisms, then there is a left $(A \otimes_R B^{op})$-module isomorphism $B_f \simeq B_g$ if and only if there exists a unit $b \in B^*$ such that*
$$g(a) = bf(a)b^{-1} \text{ for all } a \in A,$$
i.e. if and only if f and g differ by an inner automorphism of B.

Proof. "only if": call $\phi: B_f \to B_g$ the given isomorphism of $(A \otimes_R B^{op})$-modules; set $b := \phi(1)$, then $\phi(x) = \phi(\Omega_f(1 \otimes x)(1)) = \Omega_g(1 \otimes x)(\phi(1)) = bx$ for all $x \in B$. since ϕ is an isomorphism necessarily $b \in B^*$. Finally $bf(a) = \phi(f(a)) = \phi(\Omega_f(a \otimes 1)(1)) = \Omega_g(a \otimes 1)(\phi(1)) = g(a)b$ for all $a \in A$. "if": given $b \in B^*$, define $\phi: B_f \to B_g$, $x \mapsto bx$; obviously ϕ is then a Z-module isomorphism. Going through the above calculations backwards shows that ϕ is even an isomorphism of $(A \otimes_R B^{op})$-modules. □

Skolem-Noether Theorem. *Let A,B be simple rings, $K := Z(B) \subseteq Z(A)$ and $|A:K|$ finite. If $f,g: A \to B$ are K-algebra homomorphisms, then there exists a unit $b \in B^*$ such that $g(a) = bf(a)b^{-1}$ for all $a \in A$.*

Corollary 1. *Let A be a simple ring, $K := Z(A)$ and $|A:K|$ finite. Then every K-algebra automorphism of A is an inner automorphism.*

Corollary 2. *Let A,A' be simple subrings of the simple ring B,*

$K := Z(B) \subseteq Z(A) = Z(A')$ *and* $A \simeq A'$ *as K-algebras. Then this isomorphism arises from an inner automorphism of* B *if* $|A:K|$ *is finite.*

Proof. The corollaries are obvious (e.g. in Corollary 1 take $A = B$ and $f = id_A$); as for the theorem we note that (thanks to Theorem 1) it suffices to show $B_f \simeq B_g$ as left $(A \otimes_K B^{op})$-modules. By Theorem 5 in §5. in connection with Wedderburn's Main Theorem in §3. we have $A \otimes_K B^{op} \simeq M_n(D)$ with unique n and some skew field D (which is unique up to isomorphism), hence - if ℓ is a minimal left ideal of $A \otimes_K B^{op}$ - $B_f \simeq \bigoplus_{i=1}^{m} \ell$ as well as $B_g \simeq \bigoplus_{j=1}^{r} \ell$ (because of Theorem 1 in §3.; note that in §3. we dealt with right ideals, however, analogous results hold for left ideals). Since both B_f and B_g are left D-vector spaces of the same finite dimension, necessarily $m = r$ and therefore $B_f \simeq B_g$ as left $(A \otimes_K B^{op})$-modules. □

Example 1. Let D be a skew field, $K := Z(D)$ ($|D:K|$ may be infinite) and $f \in K[T]$ an irreducible polynomial, then we obtain from Corollary 2: if $d, d' \in D$ such that $f(d) = 0 = f(d')$, then $d' = bdb^{-1}$ for suitable $b \in D^*$.

Now we introduce a new notion:

Definition 2. *Let* A *be a ring,* $B \subseteq A$ *a subring and* $M \subseteq A$ *a subset; define the "centralizer of* M *in* B *" as the set*

$$Z_B(M) := \{ b \in B \mid bm = mb \text{ for all } m \in M \}.$$

In this context the following can be observed immediately:

(1) $Z_B(M)$ is a subring of B and hence also of A ;

(2) $Z_B(M) = Z_B(\langle M \rangle_A)$ where $\langle M \rangle_A$ denotes the subring of A which is generated by the set M ;

(3) $Z_B(M)$ is an R-algebra provided A and B are ;

(4) $Z_B(M) = B \cap Z_A(M)$;

(5) $Z(A) = Z_A(A)$;

(6) $Z_A(Z(A)) = A$;

(7) $M \subseteq Z_A(Z_A(M))$;

(8) $M \supseteq N$ implies $Z_B(M) \subseteq Z_B(N)$;

(9) $C \subseteq B \subseteq A$ implies $Z_C(M) \subseteq Z_B(M)$;

(10) $B \subseteq Z_A(B)$ if and only if B is commutative ;

(11) $B = Z_A(B)$ if and only if B is maximal commutative in A ;

(12)   if $f: A \to A'$ is an isomorphism, then $f(Z_A(B)) = Z_{A'}(f(B))$

Theorem 2. *Let K be a commutative field and A,A',B,B' K-algebras such that $A' \subseteq A$ and $B' \subseteq B$; then*

$$Z_{A \otimes_K B}(A' \otimes_K B') = Z_A(A') \otimes_K Z_B(B') \text{ in } A \otimes_K B.$$

Proof. First we point out that the statements of the theorem are to be understood with the conventions explained preceding Theorem 4 in §5., i.e. $A' \otimes_K B'$ resp. $Z_A(A') \otimes_K Z_B(B')$ are being regarded as K-subalgebras of the K-algebra $A \otimes_K B$. Now we note that "$\supseteq$" is clear from the various definitions; as for the converse we select bases $\{e_i\}_{i \in I}$ resp. $\{f_j\}_{j \in J}$ of A resp. B over K and choose $x \in Z_{A \otimes_K B}(A' \otimes_K B')$. It follows that there exist

$b_i \in B$ resp. $a_j \in A$ (uniquely determined by x and $= 0$ for all but finitely many indices i resp. j)

such that

$$\sum_{i \in I} e_i \otimes b_i = x = \sum_{j \in J} a_j \otimes f_j \quad \text{(cf. Theorem 3 in §4.)},$$

hence, by our assumptions on x,

$$\sum_{i \in I} e_i \otimes b_i b' = x(1 \otimes b') = (1 \otimes b')x = \sum_{i \in I} e_i \otimes b' b_i$$

for all $b' \in B$, hence (again by Theorem 3 in §4.) $b_i b' = b' b_i$ for all $b' \in B$, i.e. $b_i \in Z_B(B')$ for all $i \in I$ and similarly $a_j \in Z_A(A')$ for all $j \in J$. Therefore

$$x \in Z_A(A') \otimes_K B \cap A \otimes_K Z_B(B') = Z_A(A') \otimes_K Z_B(B'),$$

the last equality being again a consequence of Theorem 3 in §4. together with the fact that a basis of e.g. $Z_A(A')$ over K can be extended to a basis of A over K. □

Let us state a few consequences of the above:

Corollary 3. *Let K be a field and A, B K-algebras, then*

$$Z(A \otimes_K B) = Z(A) \otimes_K Z(B). \quad □$$

Now we recall Definition 3 in §5.: combining it with Theorem 5 in §5. and Corollary 3 above provides

Corollary 4. *A and B are central simple K-algebras if and only if $A \otimes_K B$ is a central simple K-algebra.* □

Another consequence of Theorem 2 is

Corollary 5. *Let* K *be a field and* A *a K-algebra, then* $Z(A) = K$ *if and only if* $Z(A \otimes_K L) = L$ *(here* L/K *is a not necessarily finite field extension)*. □

Using again Definition 3 and Theorem 5 in §5. we deduce from the previous corollary

Corollary 6. *If* L/K *is a (not necessarily finite) field extension, then* A *is a central simple K-algebra if and only if* $A \otimes_K L$ *is a central simple L-algebra*. □

Recalling Definition 4 in §6. we finally obtain (cf. Theorem 5 in §5., Theorems 1/2 in §6. and Corollary 5):

Corollary 7. *Let* L/K *be finite Galois,* $\Gamma := \text{Gal}(L/K)$ *and* A *an* L/K-*Galois algebra; then* A *is a simple ring with centre* L *if and only if* A^Γ *is a simple ring with centre* K. □

Now let R be a commutative ring and A an R-algebra; in what follows we shall deal with the R-algebra homomorphisms $L: A \to \text{End}_R(A)$, $a \mapsto L_a$ and $R: A^{op} \to \text{End}_R(A)$, $a \mapsto R_a$ known already from Lemma 5 *etc.* in §5. (here L_a resp. R_a denotes the left resp. right multiplication by a on A). With this notation we claim

Lemma 1. $Z_{\text{End}_R(A)}(L(A)) = R(A)$ *and* $Z_{\text{End}_R(A)}(R(A)) = L(A)$.

Proof. In both cases "⊇" is obviously true by the law of associativity of the multiplication in A. As for the converse let us discuss the first statement only (the second one is proved similarly): $f \in Z_{\text{End}_R(A)}(L(A))$ amounts to $fL_a = L_a f$ for all $a \in A$, hence $f(ax) = af(x)$ for all $a, x \in A$, in particular $f(a) = af(1)$ for all $a \in A$, i.e. $f = R_{f(1)} \in R(A)$. □

Now we are fully prepared to prove the

Centralizer Theorem. *Let* B *be a simple subring of a simple ring* A, $K := Z(A) \subseteq Z(B)$ *and* $n := |B:K|$ *finite, then:*
(i) $\quad Z_A(B) \otimes_K M_n(K) \simeq A \otimes_K B^{op}$;
(ii) $\quad Z_A(B)$ *is a simple ring*;
(iii) $\quad Z(Z_A(B)) = Z(B)$;
(iv) $\quad Z_A(Z_A(B)) = B$;
(v) $\quad$ *if* $L := Z(B)$ *and* $r := |L:K|$, *then* $A \otimes_K L \simeq M_r(B \otimes_L Z_A(B))$;
(vi) $\quad$ A *is a free left(right)* $Z_A(B)$-*module of unique rank* n;

(vii)   *if, in addition to the above assumptions,* $m := |A:K|$ *is also finite, then* A *is a free left(right)* B-*module of unique rank* $\frac{m}{n} = |Z_A(B):K|$.

Corollary 8. *Let* $B \subseteq A$ *be simple rings such that* $K := Z(A) = Z(B)$, *then* $A \simeq B \otimes_K Z_A(B)$ *whenever* $|B:K|$ *is finite.*

Proof. The corollary is just *(v)* in case $L = K$, i.e. $r = 1$. For the proof of the theorem consider the K-algebra homomorphisms

$$f,g: B \to A \otimes_K \text{End}_K(B) =: C \ ; \ f(b) := b \otimes \text{id}_B \ , \ g(b) := 1 \otimes L_b \ ;$$

thanks to Corollary 3 and Theorem 5 in §5. C is a simple ring with centre K , hence we can apply the Skolem-Noether Theorem which implies $f(B) \simeq g(B)$ under an inner automorphism of C . By construction of f and g we get (notation as in Lemma 1)

$$B \otimes_K K = f(B) \simeq g(B) = K \otimes_K L(B) \ ,$$

hence by (12), Theorem 2 and Lemma 1

(13) $Z_A(B) \otimes_K \text{End}_K(B) = Z_C(f(B)) \simeq Z_C(g(B)) = A \otimes_K R(B)$.

Now apply (12), Corollary 3 (resp. Theorem 2) and Lemma 1 to the previous result; this gives (note $\text{End}_K(B) \simeq M_n(K)$ and - since B has no proper two-sided ideals - $R(B) \simeq B^{op}$)

(14) $Z(Z_A(B)) \otimes_K K = Z(Z_C(f(B))) \simeq Z(Z_C(g(B))) = K \otimes_K Z(B^{op})$

resp.

(15) $Z_A(Z_A(B)) \otimes_K K = Z_C(Z_C(f(B))) \simeq Z_C(Z_C(g(B))) = K \otimes_K L(B)$.

Now (13) amounts to *(i)* (cf. the note previous to (14)) and *(i)* implies *(ii)* thanks to Theorem 5 in §5. (14) resp. (15) imply

$$Z(Z_A(B)) \simeq Z(B) \quad \text{resp.} \quad Z_A(Z_A(B)) \simeq B \quad (\text{note } L(B) \simeq B) \ ,$$

hence even "=" in both cases since we have obviously "⊇" in both cases together with the fact that on both sides in either inequality there are isomorphic finite dimensional vector spaces over K . This completes the proof of *(iii)* and *(iv)*. Now we make use of *(i)* and the various isomorphisms of §§4/5 which provides

$$Z_A(B) \otimes_L M_n(L) \simeq Z_A(B) \otimes_L (L \otimes_K M_n(K)) \simeq Z_A(B) \otimes_K M_n(K) \simeq$$
$$\simeq A \otimes_K B^{op} \simeq (A \otimes_K L) \otimes_L B^{op}$$

and therefore (use Corollaries 1/2 in §5.) - if $n = rs$ -

$$(B \otimes_L Z_A(B)) \otimes_L (M_r(L) \otimes_L M_s(L)) \simeq (A \otimes_K L) \otimes_L (B^{op} \otimes_L B) \simeq$$
$$\simeq (A \otimes_K L) \otimes_L M_s(L) \ .$$

Now we make use of Lemma 4 in §5. and get

$$M_s(A \otimes_K L) \simeq M_s(M_r(B \otimes_L Z_A(B)))$$

which implies *(v)* by Wedderburn's Main Theorem in §3. (note that $A \otimes_K L$ and $M_r(B \otimes_L Z_A(B))$ are simple rings with centre L). Since *(vii)* is a consequence of *(i),..,(vi)* (because the finiteness of $|A:K|$ implies the finiteness of $|Z_A(B):K|$ which means that we may interchange the roles of B and $Z_A(B)$ in *(i),..,(vi)*) we are left with the proof of *(vi)*: but *(vi)* is an immediate consequence of the following lemma (the rank being calculated with the aid of *(i)* together with Theorem 3 in §4.):

Lemma 2 . *Let A be a simple ring which is a subring of a ring R , then R is a free right(left) A-module of unique [possibly infinite] right (left) rank.*

Proof. Let $\hbar$ be a minimal right ideal of A , then - by Theorem 1 in §3.- $A \simeq \bigoplus_{i=1}^{n} \hbar$ as a right A-module. Since R is an A-bimodule (via the multiplication in R) we may view $\hbar \otimes_A R$ as a right A-module (cf. Lemma 3 in §4.) and obtain the right A-module homomorphism

$$\hbar \otimes_A R \simeq \bigoplus_{j \in J} \hbar \; ;$$

the proof thereof coincides with the one of Theorem 1 in §3., except for the fact that the set J is not necessarily finite because we have no finiteness condition - such as e.g. the Artinian condition - available. It follows easily (cf. Theorems 2/3 in §4.)

$$R \simeq A \otimes_A R \simeq (\bigoplus_{i=1}^{n} \hbar) \otimes_A R \simeq \bigoplus_{i=1}^{n} (\hbar \otimes_A R) \simeq \bigoplus_{i=1}^{n} \bigoplus_{j \in J} \hbar \simeq$$

$$\simeq \bigoplus_{j \in J} \bigoplus_{i=1}^{n} \hbar \simeq \bigoplus_{j \in J} A \quad \text{as right A-modules,}$$

i.e. R is free as asserted (similarly for left modules). It remains to be shown that the (possibly infinite) rank is unique; this, however, is clear since A is a finite dimensional vector space over a skew field (cf. Wedderburn's Main Theorem in §3.). ▫ ▫

In practice we shall frequently make use of

Lemma 3 . *Let B be a simple subring of a skew field D , then B and $Z_D(B)$ are also skew fields.*

Proof. B is a simple ring without zero-divisors $\neq 0$, hence a skew field thanks to Theorem 3 in §3. Consequently $Z_D(B)$ is also a skew field, since the centralizer of a skew field in a skew field is clearly a skew field. □

Definition 3 . *In the situation (and with the notation) of* Lemma 2 *one calls the right(left) rank of* R *the "right(left) degree of* R *over* A *" and denotes it* $|R:A|_R$ *(*$|R:A|_L$*)*.

Now, in precisely the same way as in the case of field extension degrees, we can prove the *degree tower formula*

Lemma 4 . *If* A,B *are simple subrings of a ring* R *such that* A ⊆ B , *then* $|R:B|_R |B:A|_R = |R:A|_R$ *(and similarly for left degrees).* □

Note that *(vi)* and *(vii)* in the Centralizer Theorem imply that certain left degrees are finite and equal to the corresponding right degrees. Even in the skew field case (cf. Lemma 3) this is not always true: one can construct examples where one of the two degrees is finite and the other infinite (see section 5.6 in P.M. Cohn [1977]), but no examples are known where both degrees are finite and different.

Theorem 3 . *Let* L *be a commutative subfield of a simple ring* A , K := := Z(A) ⊆ L *and* $|L:K|$ *finite, then we have* L = $Z_A(L)$ *if and only if* $|L:K|^2 = |A:K|$.

Proof. Since L is simple the "only if" is an easy consequence of Lemma 4 and *(vi)* in the Centralizer Theorem (take B = L). Conversely, (10) implies L ⊆ $Z_A(L)$, hence "=" thanks to *(vii)* in the Centralizer Theorem (take B = L) for dimensional reasons. □

Now recall Definition 3 in §5.; in this sense we claim

Theorem 4 . *Let* D *be a* K-*skew field, then it posesses maximal commutative subfields and all of these include* K . *Moreover,* L *is a maximal commutative subfield of* D *if and only if* $|L:K|^2 = |D:K|$.

Proof. The first two statements are clear, the "if" follows from Theorem 3 together with (11) and the "only if" holds for the same reasons plus the fact that a maximal commutative subring in a skew field must be a field. □

The next results deal with refinements of the Centralizer Theorem in certain special cases.

Theorem 5 . *Let* A *be a simple ring,* L *a commutative subfield of* A *such that* K := Z(A) ⊆ L ⊆ A , n := $|L:K|$ *and* B := $Z_A(L)$; *then* L/K *is a Galois extension if and only if there exist elements* e_i ∈ A* *such* that $Be_i = e_iB$ ($i = 1,..,n$) and $\bigoplus_{i=1}^{n} Be_i = A$ ($e_1 = 1$).

Proof. "if": by assumption we have $e_iBe_i^{-1} = B$, hence - since $L = Z(B)$ thanks to (iii) in the Centralizer Theorem - $\sigma_i(x) := e_i x e_i^{-1}$ ($x \in L$) defines K-algebra automorphisms of L ($i = 1,..,n$). If now $\sigma_i = \sigma_j$ for some indices $i \neq j$, then $e_i x e_i^{-1} = e_j x e_j^{-1}$ for all $x \in L$, i.e. $e_j^{-1} e_i \in Z_A(L) = B$, hence $e_i \in e_j B$ which contradicts the left linear independence of the e_i over B. Therefore $\Gamma := \{\sigma_1,..,\sigma_n\}$ constitutes a set of n distinct K-algebra automorphisms of L ; this means that L/K is Galois (because of $n = |L:K|$) with $\Gamma = Gal(L/K)$. "only if": let $\Gamma := Gal(L/K)$, hence $|\Gamma| = |L:K| = n$. Using the Skolem-Noether Theorem, we may find for every $\sigma \in \Gamma$ an element $e_\sigma \in A^*$ such that

$$\sigma(x) = e_\sigma x e_\sigma^{-1} \text{ for all } x \in L ,$$

hence (thanks to (12)) $e_\sigma B e_\sigma^{-1} = B$ and therefore $Be_\sigma = e_\sigma B$ ($\sigma \in \Gamma$). Thanks to $e_\sigma \in A^*$ any single one of the e_σ is left linearly independent over B . Now suppose that any $t-1$ of the elements e_σ are left linearly independent ($1 < t \leq n = |\Gamma|$) and let T be a subset of Γ with exactly t elements; fix an element $\tau_0 \in T$ and choose $x \in L$ such that $\tau_0(x) \neq \tau(x)$ for all $\tau \in T$, $\tau \neq \tau_0$. If then

$$\sum_{\tau \in T} b_\tau e_\tau = 0 ,$$

we get - recall $L = Z(B)$ -

$$\sum_{\tau_0 \neq \tau \in T} b_\tau e_\tau x e_{\tau_0}^{-1} e_{\tau_0} = \sum_{\tau_0 \neq \tau \in T} e_{\tau_0} x e_{\tau_0}^{-1} b_\tau e_\tau = -e_{\tau_0} x e_{\tau_0}^{-1} b_{\tau_0} e_{\tau_0} =$$

$$= -b_{\tau_0} e_{\tau_0} x = \sum_{\tau_0 \neq \tau \in T} b_\tau e_\tau x = \sum_{\tau_0 \neq \tau \in T} b_\tau e_\tau x e_\tau^{-1} e_\tau ,$$

therefore (by induction assumption) $b_\tau(\tau_0(x) - \tau(x)) = 0$ for all $\tau \in T$, $\tau \neq \tau_0$. By our choice of x the latter amounts to $b_\tau = 0$ for these τ , hence also $b_{\tau_0} e_{\tau_0} = 0$, i.e. even $b_\tau = 0$ for all $\tau \in T$. Consequently any t elements of the e_σ are left linearly independent, therefore, by induction, all n of these are. Finally

$$A = \bigoplus_{\sigma \in \Gamma} Be_\sigma ,$$

because the span on the right hand side is a finite dimensional left vector space over some skew field of the same dimension as A has (cf. (vi) in the Centralizer Theorem). Now set $e_i := e_{\sigma_i}$ if $\Gamma = \{1 = \sigma_1, \sigma_2,...,\sigma_n\}$. □

Inspection of the above proof gives immediately

Theorem 6. *Let* A *be a simple ring,* L *a commutative subfield of* A *such that* $K := Z(A) \subseteq L \subseteq A$, $n := |L:K|$ *and* $B := Z_A(L)$; *then* L/K *is a cyclic extension if and only if there exists an element* $e \in A^*$ *such that* $Be = eB$ *and* $\bigoplus_{i=0}^{n-1} Be^i = A$. □

We close this paragraph with a theorem due to O. Teichmüller [1940] which is, in some sense, a converse to parts of the previous theorem.

Theorem 7. *Let* L/K *be a finite cyclic extension,* $\Gamma := \text{Gal}(L/K) = \langle \sigma \rangle$ *and* B *a simple ring with centre* L. *If* σ *can be prolonged to a ring automorphism* Φ *of* B, *then there is a simple ring* A *with centre* K *such that* $B = Z_A(L)$ *and hence* $A \otimes_K L \simeq M_n(B)$ ($n := |L:K|$).

Proof. The last claim follows from the second last one thanks to *(v)* in the Centralizer Theorem. Now the actual proof: because of $\Phi^n|_L = \sigma^n = \text{id}_L$ the Skolem-Noether Theorem provides the existence of an element $a_0 \in B^*$ such that

(16) $\qquad \Phi^n(b) = a_0 b a_0^{-1}$ for all $b \in B$;

of course, a_0 in (16) is uniquely determined by Φ only up to multiplication with elements from L^* (note $L = Z(B)$). It follows

$$a_0 \Phi(b) a_0^{-1} = \Phi^{n+1}(b) = \Phi(a_0 b a_0^{-1}) = \Phi(a_0)\Phi(b)\Phi(a_0)^{-1}$$

for all $b \in B$, therefore

$$t_0 := \Phi(a_0)^{-1} a_0 \in Z(B)^* = L^*$$

and consequently $\Phi(a_0)^{-1} t_0 a_0 = t_0 \Phi(a_0)^{-1} a_0 = \Phi(a_0)^{-1}(a_0 \Phi(a_0)^{-1}) a_0$, hence even

$$\Phi(a_0)^{-1} a_0 = t_0 = a_0 \Phi(a_0)^{-1}$$

and thus

$$N_{L/K}(t_0) = \prod_{i=0}^{n-1} \sigma^i t_0 = \prod_{i=0}^{n-1} (\Phi^{i+1}(a_0)^{-1} \Phi^i(a_0)) = 1 .$$

Now Hilbert's "Satz 90" in §6. provides the existence of an element $t \in L^*$ such that

$$\Phi(a_0)^{-1} a_0 = t_0 = {}^\sigma t t^{-1} = \sigma(t) t^{-1} = \Phi(t) t^{-1} ,$$

so if we replace in (16) our element a_0 by $a := a_0 t$ we get $\Phi(a) = a$, in other words we can even find an element $a \in B^*$ such that

(17) $\qquad \Phi(a) = a$ and $\Phi^n(b) = aba^{-1}$ for all $b \in B$.

Therefore, to complete our proof, it suffices to prove the following

Lemma 5. *Let L/K be a cyclic extension, $n := |L:K|$, $\Gamma := \mathrm{Gal}(L/K) =$
$= \langle \sigma \rangle$, B a simple ring with centre L, Φ a ring automorphism of B
such that $\Phi|_L = \sigma$ and finally a an element of B^* such that (17)
holds; then*

(18) $\qquad A := \bigoplus_{i=0}^{n-1} B e^i$ *with multiplication* $e^n = a$, $eb = \Phi(b)e$

*defines a simple ring with centre K such that $B = Z_A(L)$ and hence
$A \otimes_K L \simeq M_n(B)$.*

Proof. Again, the last claim follows from the second last one (cf. the proof of the previous theorem). Now an entirely straightforward calculation shows that (in view of (17)) the multiplication defined in (18) is reasonable in the sense that A is endowed with the structure of a ring with unit element $e^0 =: 1$. B (and therefore L (note $L \subseteq B$)) may be viewed being embedded into A via $b \mapsto be^0 = b1$; in this sense $B \subseteq Z_A(L)$ is clear. We want to prove the converse: let $z = \sum_{i=0}^{n-1} b_i e^i \in Z_A(L)$ be given (with uniquely determined $b_i \in B$) and choose $x \in L$ such that $\sigma^i x \neq x$ for all $i = 1,\ldots,n-1$. It follows

$$\sum_{i=0}^{n-1} x b_i e^i = xz = zx = \sum_{i=0}^{n-1} b_i e^i x = \sum_{i=0}^{n-1} b_i \Phi^i(x) e^i = \sum_{i=0}^{n-1} \sigma^i x b_i e^i$$

which amounts to $b_i = 0$ for $i = 1,\ldots,n-1$ (by our choice of x), hence $z = b_0 e^0 = b_0 1 \in B$, i.e. $B = Z_A(L)$. From the latter we get

$$K \subseteq Z(A) = Z_A(A) \subseteq Z_A(L) = B, \text{ hence } Z(A) \subseteq Z(B) = L$$

and even $K = Z(A)$, since $xe = ex = \Phi(x)e = \sigma xe$ implies $x \in \mathrm{Fix}_L(\Gamma) = K$. By virtue of Corollary 4 in §5., in order to complete the proof of our lemma, we are left with showing that A according to (18) has no proper two-sided ideal; indeed, let $0 \neq y \in A$, then $y = \sum_{i \in Y} b_i e^i$ with some well-defined non-empty subset $Y = Y(y) \subseteq \{0,1,\ldots,n-1\}$ such that $b_i = 0$ for all indices $i \notin Y$. If now $\{0\} \neq \mathfrak{a} \subseteq A$ is a two-sided ideal of A and if $|Y(y)| = 1$ for some $0 \neq y \in \mathfrak{a}$, it follows $y = b_j e^j$ for some j, i.e. $1 = b_j^{-1} y (e^j)^{-1} \in \mathfrak{a}$ and hence $\mathfrak{a} = A$ (note that (18) implies $(e^i)^{-1} = a^{-1} e^{n-i}$). If on the other hand $|Y(y)| \geq 2$, select $x \in L$ and $j,r \in Y = Y(y)$ such that $\Phi^j(x) = \sigma^j x \neq \sigma^r x = \Phi^r(x)$, and consider

$$\mathfrak{a} \ni y - \Phi^j(x)^{-1} yx = \sum_{i \in Y} (b_i - \Phi^j(x)^{-1} \Phi^i(x) b_i) e^i =$$
$$=: \sum_{i \in Y'} b_i' e^i =: y'$$

with $b_i' \in B$, $b_r \neq 0$ (by our choice of x) and $Y' := Y \smallsetminus \{j\} \supseteq Y(y')$. Because of $|Y(y')| \leq |Y'| < |Y|$ we may conclude $a = A$ by induction; this completes the proof of the lemma as well as the one of the theorem previous to the lemma. □ □

Lemma 5 is already remarkable in the (seemingly easy) special case $B = L$ and gives rise to the following

Definition 4. Let L/K *be a cyclic extension*, $n := |L:K|$, $\Gamma := \mathrm{Gal}(L/K) = \langle \sigma \rangle$ *and* $a \in K^*$, *then one calls the ring*
$$(a, L/K, \sigma) := \bigoplus_{i=0}^{n-1} Le^i \text{ with multiplication } e^n = a, \text{ } ex = {}^\sigma xe$$
($x \in L$) *a "cyclic algebra"*.

Lemma 5 reads then as follows:

Lemma 6. *Let* $A := (a, L/K, \sigma)$ *be a cyclic algebra, then* A *is a central simple* K-*algebra such that* $L = Z_A(L)$ *and* $A \otimes_K L \simeq M_n(L)$. □

Exercise 1. Let A be a simple ring with centre K and B a subring of A containing K such that A is a finitely generated right(left) B-module. Show that $|Z_A(B):K| \leq n$ where n denotes the number of generators needed (cf. Theorem 16 in E. Artin & G. Whaples [1943] or Theorem 7.3H in E. Artin et al. [1948]).

Exercise 2. Show that the Skolem-Noether Theorem does not hold in skew fields which are infinite dimensional over their centre (Hint. Set $L := \bigl(K((X))\bigr)((Y))$, define an automorphism σ of L such that $\sigma(X) = X+X^2$ and such that $D := L((T;\sigma))$ has centre K (cf. §1.); now define a K-algebra automorphism τ of D suitably and prove that it cannot be an inner automorphism.).

Exercise 3. Modify the constructions from Exercise 2 in such a way that you can show that e.g. *(iii)* and *(iv)* in the Centralizer Theorem do not hold in case $|B:K|$ is infinite.

§ 8 . THE CORESTRICTION OF ALGEBRAS

Let R be a *commutative* ring and A an R-algebra, i.e. a ring with $1 \neq 0$ together with a fixed ring homomorphism $i_A: R \to Z(A) \subseteq A$. Now let σ be a ring automorphism of R; with the aid of σ let us equip our *ring* A with a *new algebra structure*:

Definition 1. *Let A be an R-algebra and σ a ring automorphism of R; define the R-algebra $^\sigma A$ to be the ring A equipped with the new R-structure $i_A \sigma^{-1}: R \to Z(A) \subseteq A$ (where i_A defines the old R-structure on A).*

The following is then quite obvious (cf. the beginning of §5.):

Lemma 1. *Let A,B be R-algebras, $f: A \to B$ a ring homomorphism and σ a ring automorphism of R; then $f: A \to {}^\sigma B$ is an R-algebra homomorphism if and only if $f i_A = i_B \sigma^{-1}$, i.e. if and only if the diagram shown commutes.*

$$\begin{array}{ccc} A & \xrightarrow{f} & B \\ {\scriptstyle i_A}\uparrow & & \uparrow{\scriptstyle i_B} \\ R & \xrightarrow{\sigma^{-1}} & R \end{array}$$ □

Of course, the commutativity of the diagram above reads in more conventional notation (writing as usual ra in place of $i_A(r)a$):

$$f(ra) = \sigma^{-1}(r)f(a) \quad (r \in R, a \in A),$$

i.e. f is semilinear (in the usual sense). Moreover we find easily

Lemma 2. $A \simeq {}^\sigma A$ *as an R-algebra if and only if σ^{-1} (and hence σ) can be extended to a ring automorphism of A.* □

In this context the following can be observed immediately:

(1) $\quad \mathrm{id}_A = A \quad \text{and} \quad {}^\sigma({}^\tau A) = {}^{\sigma\tau}A$;

(2) $\quad {}^\sigma R \simeq R$ as an R-algebra ;

(3) $\quad Z({}^\sigma A) = Z(A)$;

(4) $\quad {}^\sigma(M_n(A)) = M_n({}^\sigma A)$;

(5) $\quad {}^\sigma(A \otimes_R B) = {}^\sigma A \otimes_R {}^\sigma B$.

Furthermore we claim

(6) If $S \subseteq \text{Fix}_R(\sigma)$ is a subring, A an S-algebra and B an R-algebra, then $^\sigma(A \otimes_S B) = A \otimes_S {}^\sigma B$ as an R-algebra.

Indeed, $A \otimes_S {}^\sigma B$ resp. $A \otimes_S B$ carry an R-algebra structure via $r \mapsto 1 \otimes i_B \sigma^{-1}(r)$ resp. $r \mapsto 1 \otimes i_B(r)$, hence the identity $f: A \otimes_S {}^\sigma B \to A \otimes_S B$ fulfills the requirements of Lemma 1. Combination of (6) and (2) yields

(7) If $S \subseteq \text{Fix}_R(\sigma)$ is a subring and A an S-algebra, then $^\sigma(A \otimes_S R) \simeq A \otimes_S R$ as an R-algebra.

Finally we point out the obvious fact

(8) $^\sigma A$ is a simple ring if and only A is a simple ring.

From now on we shall deal only with the case where R is a field, more precisely: let N/K be a finite Galois extension, L an intermediate field, $\Delta := \text{Gal}(N/L) \leq \text{Gal}(N/K) =: \Gamma$, R a system of representatives for the cosets of Γ modulo Δ - i.e. $\Gamma = \bigcup_{\rho \in R} \rho\Delta$ - and A an L-algebra; in this situation we claim (and this will be of great importance in the sequel):

Lemma 3. *Given* $\rho \in R$, *then* $^\rho(A \otimes_L N)$ *is (up to N-algebra isomorphism) independent of the particular choice of the system* R, *hence the same is true for the N-algebra* $A^{(\Gamma:\Delta)} := \bigotimes_{\rho \in R} {}^\rho(A \otimes_L N)$.

Proof. Thanks to (1) and (7) we have for all $\delta \in \Delta$ the N-algebra isomorphisms $^{\rho\delta}(A \otimes_L N) \simeq {}^\rho({}^\delta(A \otimes_L N)) \simeq {}^\rho(A \otimes_L N)$. □

Now let R be as above; given $\rho \in R$ and $\sigma \in \Gamma$, there are (uniquely determined) elements

(9) $^\sigma\rho \in R$ and $\delta(\sigma,\rho) \in \Delta$ such that $\sigma\rho = {}^\sigma\rho\,\delta(\sigma,\rho)$.

It follows ($\sigma, \tau \in \Gamma$; $\rho \in R$)

$$\sigma\tau\rho\,\delta(\sigma\tau,\rho) = (\sigma\tau)\rho = \sigma(\tau\rho) = \sigma{}^\tau\rho\,\delta(\tau,\rho) = {}^\sigma({}^\tau\rho)\delta(\sigma,{}^\tau\rho)\delta(\tau,\rho),$$

hence

(10) $^{\sigma\tau}\rho = {}^\sigma({}^\tau\rho)$, $^1\rho = \rho$ and $\delta(\sigma\tau,\rho) = \delta(\sigma,{}^\tau\rho)\delta(\tau,\rho)$,
$\delta(1,\rho) = 1$.

Using the notions and notations just introduced we claim:

Lemma 4. *Consider the N-algebra* $A^{(\Gamma:\Delta)}$ *(as defined in Lemma 3), then* $A^{(\Gamma:\Delta)}$ *can be given the structure of an N/K-Galois algebra* (cf. Definition 4 in §6.) *such that*

(11) $\quad {}^\sigma(..\otimes a_\rho \otimes..) = ..\otimes^{\delta(\sigma,\rho)} a_\sigma \otimes..\quad$ ($\sigma \in \Gamma$, $a_\rho \in {}^\rho(A \otimes_L N)$)

$\quad\quad\quad\quad\quad\quad\quad\uparrow\rho$-th position $\quad\uparrow\sigma\rho$-th position

where $A \otimes_L N$ *is viewed as a left* Δ-*module via* $\mathrm{id}_A \otimes \sigma$ ($\sigma \in \Delta$).

Proof. It follows immediately from Theorem 8 in §4. that $x \mapsto {}^\sigma x$ (according to (11)) defines a ring automorphism of $A^{(\Gamma:\Delta)}$ for every $\sigma \in \Gamma$. Since ${}^1 x = x$ is obvious from (10) we must now show ${}^\sigma({}^\tau x) = {}^{\sigma\tau} x$ ($\sigma, \tau \in \Gamma$). Again this is clear from (10) thanks to

$\quad\quad\quad\quad\quad\quad\uparrow\rho$-th position $\quad\uparrow\sigma\tau\rho$-th position

$${}^{\sigma\tau}(..\otimes a_\rho \otimes..) = ..\otimes^{\delta(\sigma\tau,\rho)} a_\rho \otimes.. = ..\otimes^{\delta(\sigma,{}^\tau\!\rho)\delta(\tau,\rho)} a_\rho \otimes.. =$$

$$= {}^\sigma(..\otimes^{\delta(\tau,\rho)} a_\rho \otimes..) = {}^\sigma({}^\tau(..\otimes a_\rho \otimes..))$$

$\quad\quad\quad\quad\quad\quad\uparrow {}^\tau\rho$-th pos. $\quad\quad\uparrow\rho$-th position .

Finally we see

$\quad\quad\quad\quad\quad\quad\uparrow\rho$-th position $\quad\quad\uparrow\sigma\rho$-th position

$${}^\sigma x^\sigma(..\otimes a_\rho \otimes..) = {}^\sigma x(..\otimes^{\delta(\sigma,\rho)} a_\rho \otimes..) = ..\otimes^{\sigma\rho^{-1}} x^{\delta(\sigma,\rho)} a_\rho \otimes.. =$$

$$= ..\otimes^{\delta(\sigma,\rho)}({}^{\rho^{-1}} x a_\rho) \otimes.. = {}^\sigma(..\otimes^{\rho^{-1}} x a_\rho \otimes..) = {}^\sigma(x(..\otimes a_\rho \otimes..))$$

$\uparrow\sigma\rho$-th position $\quad\quad\quad\quad\quad\quad\quad\quad\uparrow\rho$-th position

which completes our proof. □

If we apply Theorems 1/2 in §6. we get from the previous result

Corollary 1 . $\left(A^{(\Gamma:\Delta)}\right)^\Gamma$ *is a K-algebra such that* $\left(A^{(\Gamma:\Delta)}\right)^\Gamma \otimes_K N \simeq$
$\simeq A^{(\Gamma:\Delta)}$. □

We want to show that the K-algebra $\left(A^{(\Gamma:\Delta)}\right)^\Gamma$ is (up to isomorphism) independent of the field N, i.e. depends only on the (separable) field extension L/K. For this, however, we need more preparation:

Let M/K be a further finite Galois extension with intermediate field L and assume $N \subseteq M$ (otherwise go over to MN and apply the next lemma twice); set $H := \mathrm{Gal}(M/L) \leq \mathrm{Gal}(M/K) =: G$ and $N := \mathrm{Gal}(M/N) \trianglelefteq G$, hence $\Gamma = G/N$ and $\Delta = H/N$. Furthermore, if R is a system of representatives for the cosets of G modulo H - i.e. $G = \bigcup_{r \in R} rH$ - then the set $\{ \rho \in \Gamma \mid \rho = r|_N = rN \in G/N = \Gamma , r \in R \}$ (which we shall also denote R) is likewise a system of representatives for the cosets of Γ modulo Δ . Now consider the M-algebra isomorphisms (use (5) together with Lemmas 2/3 in §5. $|R|$-times, i.e. $|L:K|$-times)

(12)
$$\bigotimes_{r \in R} {}^r(A \otimes_L M) \simeq \bigotimes_{r \in R} {}^r((A \otimes_L N) \otimes_N M) \simeq$$
$$\simeq (\bigotimes_{\rho \in R} {}^\rho(A \otimes_L N)) \otimes_N (\bigotimes_{r \in R} {}^r M) \ ;$$

here G and *a fortiori* N operate on the left hand side according to (11). Hitherto we have not made use of the fact that N is a *normal* subgroup of G ; now we take this into account and conclude from it immediately $sr = rh(s,r)$ for all $s \in N$ (note that $h(s,r) \in H$ is the analogue in G to the function δ according to (9)), hence even
$${}^s r = r \text{ and } h(s,r) = r^{-1}sr \in N \text{ for all } s \in N \ .$$
Therefore N operates on the right hand side of (12) componentwise on each of the factors of $\bigotimes_{r \in R} {}^r M$ and trivially (cf. (7)) on $\bigotimes_{\rho \in R} {}^\rho(A \otimes_L N)$. From the first one of these two facts we conclude (cf. Theorems 1/2 in §6. and use Theorem 2 in §4. ($|R|-1$)-times) $(\bigotimes_{r \in R} {}^r M)^N \simeq \bigotimes_{r \in R} ({}^r M)^N \simeq \bigotimes_{r \in R} N \simeq$
$\simeq N$, hence (after using Theorems 1/2 in §6. again) the formula (12) implies the K-algebra isomorphisms
$$(A^{(G:H)})^G = ((A^{(G:H)})^N)^{G/N} = (\bigotimes_{\rho \in R} {}^\rho(A \otimes_L N))^\Gamma \otimes_N N^\Gamma \simeq$$
$$\simeq (A^{(\Gamma:\Delta)})^\Gamma \otimes_K K \simeq (A^{(\Gamma:\Delta)})^\Gamma \ .$$
Consequently we have proved:

Lemma 5 . *In the situation and with the notation introduced above one has a K-algebra isomorphism* $(A^{(G:H)})^G \simeq (A^{(\Gamma:\Delta)})^\Gamma$. □

Now we drop the assumption "$N \subseteq M$" and use Lemma 5 twice (with NM in place of N and M); this gives

Theorem 1 . *Let L/K be a finite separable extension and A an L-algebra; then the K-algebra $(A^{(\Gamma:\Delta)})^\Gamma$* (according to Lemma 4 and Corollary 1) *is (up to K-algebra isomorphism) independent of the auxiliary field extension N/L and therefore depends only on A and L/K* . □

Now the following definition is feasible:

Definition 2 . *If L/K is finite separable and A an L-algebra, then the K-algebra* $c_{L/K}(A) := (A^{(\Gamma:\Delta)})^\Gamma$ *is called the "corestriction of A "*.

Lemma 6 . *Let L/K be finite separable and A an L-algebra, then: A has no proper two-sided ideals and $Z(A) = L$ if and only if $c_{L/K}(A)$ has no proper two-sided ideals and $Z(c_{L/K}(A)) = K$* .

Proof. Clear from the definition of the corestriction (as regards the centre: cf. Corollaries 3/5 in §7.; as regards the two-sided ideals: cf. Corollary 3 in §5.). □

Furthermore we conclude from Lemma 5 and Theorems 3/7 in §4. in connection with the definition of the corestriction:

Lemma 7. *Let* L/K *be finite separable and* A *an* L-*algebra, then* $|c_{L/K}(A):K| = |A:L|^{|L:K|}$ *whenever either side is finite.* □

Corollary 2. *Let* L/K *be finite separable, then* A *is a central simple* L-*algebra if and only if* $c_{L/K}(A)$ *is a central simple* K-*algebra.* □

Now we want to study the special case $A = B \otimes_K L$ with some K-algebra B, however, before we can do so we must establish another important result:

Lemma 8. *Let* L/K *be finite separable, then* $c_{L/K}(L) \simeq K$.

Proof. Using (2) and Theorem 2 in §4. several times we obtain the N-algebra isomorphisms

$$(13) \qquad N \simeq \bigotimes_{\rho \in R} N \simeq \bigotimes_{\rho \in R} {}^\rho N \simeq \bigotimes_{\rho \in R} {}^\rho(L \otimes_L N) = L^{(\Gamma:\Delta)}$$

such that (note that we identify $y \in N$ with $1 \otimes_L y \in L \otimes_L N$ in the formula below)

$$x \mapsto x(1\otimes..\otimes 1) = ..1\otimes^{\rho^{-1}} x \otimes 1.. = ..1\otimes^{\sigma\rho^{-1}} x \otimes 1.. ,$$

ρ-th position ⟶ $\sigma\rho$-th position

hence (cf. (9) and (11))

$${}^\sigma x \mapsto {}^\sigma x(1\otimes..\otimes 1) = ..1\otimes^{\sigma\rho^{-1}} ({}^\sigma x) \otimes 1.. = ..1\otimes^{\delta(\sigma,\rho)}({}^{\rho^{-1}} x) \otimes 1.. =$$
$$= {}^\sigma(..1\otimes^{\rho^{-1}} x \otimes 1..) = {}^\sigma(x(1\otimes..\otimes 1)) \quad (\sigma \in \Gamma, \rho \in R),$$

↑— ρ-th position

i.e. under the isomorphism in (13) the standard Γ-action on N is carried over to the Γ-action according to (11) on $L^{(\Gamma:\Delta)}$. Therefore we may conclude $K = N^\Gamma \simeq \left(L^{(\Gamma:\Delta)}\right)^\Gamma = c_{L/K}(L)$. □

Now we are ready for the proof of (as regards the notation: cf. Definition 3 in §4.)

Lemma 9. *Let* L/K *be finite separable and* B *a* K-*algebra, then* $c_{L/K}(B \otimes_K L) \simeq B^{\otimes d}$ ($d := |L:K|$).

Proof. Using (6), (13) and Lemmata 2/3 in §5. one obtains the N-algebra isomorphisms

(14)
$$(B \otimes_K L)^{(\Gamma:\Delta)} = \bigotimes_{\rho \in R}{}^\rho((B \otimes_K L) \otimes_L N) \simeq \bigotimes_{\rho \in R}{}^\rho(B \otimes_K N) \simeq$$
$$\simeq \bigotimes_{\rho \in R}(B \otimes_K {}^\rho N) \simeq B^{\otimes d} \otimes_K (\bigotimes_{\rho \in R}{}^\rho N) \simeq B^{\otimes d} \otimes_K N \quad .$$

Here Γ operates trivially on the left term, hence our claim thanks to Theorem 1*(I)* in §6. □

Writing $M_n(L) \simeq M_n(K) \otimes_K L$ (cf. Lemma 4 in §5.) and using Lemma 9 as well as Corollary 1 in §5. we get

Corollary 3. *Let* L/K *be finite separable and* A,B L-*algebras, then*
$$c_{L/K}(M_n(L)) \simeq M_{n^d}(K) \quad (\text{ here } d := |L:K|) . \quad \square$$

Lemma 10. *Let* L/K *be finite separable and* A,B L-*algebras, then*
$$c_{L/K}(A \otimes_L B) \simeq c_{L/K}(A) \otimes_K c_{L/K}(B) \quad .$$

Proof. Use (5) and Lemma 3 in §5.; this gives N-algebra isomorphisms
$$(A \otimes_L B)^{(\Gamma:\Delta)} = \bigotimes_{\rho \in R}{}^\rho((A \otimes_L B) \otimes_L N) \simeq$$
$$\simeq \bigotimes_{\rho \in R}{}^\rho((A \otimes_L N) \otimes_N (B \otimes_L N)) \simeq$$
$$\simeq (\bigotimes_{\rho \in R}{}^\rho(A \otimes_L N)) \otimes_N (\bigotimes_{\rho \in R}{}^\rho(B \otimes_L N)) = A^{(\Gamma:\Delta)} \otimes_N B^{(\Gamma:\Delta)}$$

such that we may apply Theorems 1/2 in §6., hence our claim. □

We close our investigations by showing that the corestriction behaves functorially in L/K .

Lemma 11. *Let* L/K *be finite separable and* A *an* L-*algebra; if* I *is an intermediate field of* L/K *, then*
$$c_{L/K}(A) \simeq c_{I/K}(c_{L/I}(A)) \quad .$$

Proof. Denote $\Sigma := \text{Gal}(N/I)$, hence $\Delta \leq \Sigma \leq \Gamma$; now write
$$\Gamma = \bigcup_{\rho' \in R'} \rho' \Sigma \; , \; \Sigma = \bigcup_{\rho'' \in R''} \rho'' \Delta \; , \text{ hence } \Gamma = \bigcup_{\rho' \in R'} \bigcup_{\rho'' \in R''} \rho' \rho'' \Delta \quad ,$$

i.e. we may take $R := \{\rho'\rho'' \mid \rho' \in R', \rho'' \in R''\}$. Now use Corollary 1 and (5) several times; this gives the K-algebra isomorphisms
$$c_{I/K}(c_{L/I}(A)) = \Big(\bigotimes_{\rho' \in R'}{}^{\rho'}(c_{L/I}(A) \otimes_I N)\Big)^\Gamma \simeq$$
$$\simeq \Big(\bigotimes_{\rho' \in R'}{}^{\rho'}(\bigotimes_{\rho'' \in R''}{}^{\rho''}(A \otimes_L N))\Big)^\Gamma \simeq \Big(\bigotimes_{\rho' \in R'}\bigotimes_{\rho'' \in R''}{}^{\rho'\rho''}(A \otimes_L N)\Big)^\Gamma =$$

$$= c_{L/K}(A) \,. \quad \square$$

Exercise 1. Let L/K be finite separable and F/K an arbitrary extension such that $E := L \otimes_K F$ is a field. Show that for L-algebras A we have an F-algebra isomorphism

$$c_{L/K}(A) \otimes_K F \simeq c_{E/F}(A \otimes_L E) \,.$$

PART II . SKEW FIELDS AND BRAUER GROUPS

The history of Brauer groups began in 1929 when R. Brauer (1902-1977) proved that the set of those (isomorphism classes of) skew fields which are finite dimensional over their common centre K can be endowed with the structure of an abelian torsion group. This group is nowadays called the Brauer group Br(K) and turns out to be a subtle invariant of K which is closely related to Galois Cohomology via the crossed products ("verschränkte Produkte" in the original German terminology),as was first shown by E. Noether (1882-1935). Of course, in those days there was no Galois Cohomology, so what she suggested was the heart of a theory which was to become Galois Cohomology after general (co)homology theories had been developed later in the 1940's. Meanwhile, in the 1930's A.A. Albert (1906-1972), R. Brauer, E. Noether and H. Hasse (1898--1979) gave a comprehensive treatment which culminated in a complete determination of Br(K) in the case where K is an algebraic number field. All this is summed up in the two reports M. Deuring [1935] and A.A. Albert [1939] which are even today outstanding reading matter.

Here (in Part II of these lectures) we attempt to give a comprehensive treatment of the *basic algebraic aspects* of a modern theory of Brauer groups over fields. (Note that today there exists a corresponding theory of Brauer groups over commutative rings which we disregard in these lectures; cf. some remarks in §18.) Of course we can only present selected topics (for otherwise these lectures would be three times as long), and it is our hope that this selection will, as well as stimulating the reader's interest, also prepare him for further reading.

In §9. we begin with the definition of Br(K) and its immediate consequences; a good knowledge of Part I is needed and in particular we make full use of C. Riehm's corestriction procedure from §8. In §10. we study a special class of algebras and skew fields related to cyclic field extensions; much of the material in this paragraph has been presented before 1929, mainly by L.E. Dickson (1874-1954). Some of the results in §10. are *really classical*: for instance Frobenius' Theorem (G.F. Frobenius (1849--1917)) and Wedderburn's Theorem from 1905 (J.H.M. Wedderburn (1882-1948)). In §11. we deal with power norm residue algebras: these originate from Class Field Theory; however, following J. Milnor [1971] we present only the algebraic aspect thereof. §§12/13. contain a description of the above mentioned relationship between Brauer groups and Galois Cohomology (here we follow E. Artin *et al.* [1948]). §14. deals with quaternion algebras and skew fields (cf. also §1.); here they appear as special cases of the algebras in §11. In §15. we present the theory of Br(K) in the case where char(K) ≠ 0 . Here we combine the advantages of A.A. Albert's approach with those of O. Teichmüller's (1913-1943) and E. Witt's. In §16. we give an introduction to W. Scharlau's [1975] version of the theory of skew fields with involution. In §17. we attempt to describe the connection between Br(K) and K_2-Theory which arises from §11. Here the situation is in flux: A.S. Merkur'ev and A.A. Suslin (А.С. Меркурев & А.А. Суслин [a],[b]) have very recently achieved a spectacular result in this field ! Here we can only describe this result; the proof would comprise another book. Finally §18. contains a survey of some further results as well as suggestions for further reading.

§ 9 . BRAUER GROUPS OVER FIELDS

Henceforth (i.e. throughout Part II) we shall be dealing with *central simple K-algebras* and *K-skew fields* only (cf. Definition 3 in §5.); these are *a fortiori* simple rings and hence we may (and shall) use the theory developed in Part I. We start our investigations with

Theorem 1 . *The following statements are equivalent:*

(a)   A *is a central simple K-algebra ;*

(b)   A *is a finite dimensional K-algebra without proper two-sided ideals and such that* $K = Z(A)$;

(c)   A *is a simple ring which is finite dimensional over* $K = Z(A)$;

(d) $A \otimes_K L$ *is a central simple L-algebra (* L/K *any (not necessarily finite) field extension) ;*

(e) $A \otimes_K \overline{K} \simeq M_n(\overline{K})$ *for any algebraic closure* $\overline{K}$ *of* K *(then* $n^2 = |A:K|$ *) ;*

(f) $A \simeq M_r(D) \simeq D \otimes_K M_r(K)$ *with unique* r *and an up to isomorphism unique K-skew field* D *(then* $r^2|D:K| = |A:K|$ *).*

Moreover, A,B *are central simple K-algebras if and only if* $A \otimes_K B$ *is a central simple K-algebra.*

Proof. This is merely a summary of various statements from §§3/5/7. □

Theorem 1 gives rise to

Definition 1 . *Let* A *be a central simple K-algebra; then the skew field* D *according to* Theorem 1(f) *is called the "skew field component of* A " *and* n *according to* Theorem 1(e) *is called the "reduced degree of* A " ; *the reduced degree of the skew field component of* A *is called the "index of* A " *and is denoted by* i(A) .

Obviously we have (because this amounts to the case $r = 1$ in Theorem 1(f))

(1)   A central simple K-algebra A is a K-skew field if and only if $i(A)^2 = |A:K|$.

We list a few features of central simple K-algebras which we shall need frequently.

Lemma 1. *Let A,C be finite dimensional K-algebras such that $|C:K| \leq |A:K|$ and let $f: A \to C$ be a K-algebra homomorphism; then, if A is a central simple K-algebra, f is an isomorphism.*

Proof. Since Ker f is a two-sided ideal $\neq A$ we see that f must be injective, hence even surjective for dimensional reasons. □

Combining Lemma 1 with Theorem 2 in §5. yields

Lemma 2. *Let A,B,C be finite dimensional K-algebras such that $|C:K| \leq |A:K||B:K|$ and let $f: A \to C$, $g: B \to C$ be K-algebra homomophisms. Then $A \otimes_K B \simeq C$ provided A,B are central simple K-algebras.* □

The following is crucial for the entire Part II :

Definition 2. *Two central simple K-algebras A,B are called "similar" (write " $A \sim B$ ") if there are $s, t \in \mathbb{N}$ such that*
$$A \otimes_K M_s(K) \simeq B \otimes_K M_t(K) .$$

Again we get from §§3/4/5/7 (details are left to the reader)

Lemma 3. *Let A,B be central simple K-algebras with corresponding skew field components D,E, then:*

(α) $A \sim B$ *if and only if* $D \simeq E$;

(β) $A \sim B$ *and* $|A:K| = |B:K|$ *if and only if* $A \simeq B$;

(γ)   *"$\sim$" is an equivalence relation ;*

(δ)   *in every equivalence class modulo "$\sim$" there is (up to isomorphism) exactly one K-skew field, namely the (common) skew field component of all the members of this class.* □

Another immediate consequence of the various preceding results is (cf. in particular Corollary 2 in §5.):

Theorem 2. *Denote by $[A]$ the equivalence class containing A of central simple K-algebras modulo "$\sim$", and write $Br(K)$ for the set of these equivalence classes (for fixed base field K); then the definition*
$$[A] + [B] := [A \otimes_K B] \quad ([A],[B] \in Br(K))$$
is feasible and endows $Br(K)$ with the structure of an abelian group (Z-module) such that
$$0 = [K] = [M_n(K)] \quad \text{and} \quad -[A] = [A^{op}] .$$
□

Definition 3. *The group* $Br(K)$ *from Theorem 2 is called the "Brauer group of* K *".*

Roughly speaking $Br(K)$ is the set of K-skew fields (up to isomorphism) together with a Z-module structure such that the (trivial) skew field K is the neutral element. Moreover, it is clear from our definitions that

(2) $\qquad i(A) = i(B)$ if $[A] = [B]$.

Note that we may (and henceforth shall) replace the somewhat lengthy phrase " A is a central simple K-algebra" by " $[A] \in Br(K)$ " !

If we state Lemmas 2/3/4 in §5. in terms of our new phrasing we get

Theorem 3. *Let* L/K *be a (not necessarily finite) field extension, then the assignment* $A \mapsto A \otimes_K L$ *induces a Z-homomorphism* $r_{L/K}: Br(K) \to Br(L)$ *which is functorial in* L *(i.e.* $r_{M/K} = r_{M/L} r_{L/K}$ *if* $K \subseteq L \subseteq M$ *).* □

Definition 4. $Br(L/K) := \ker r_{L/K}$ *is called the "relative Brauer group of* L/K *"; if* $[A] \in Br(K)$ *and* L/K *is such that* $[A] \in Br(L/K)$ *(the latter amounts to* $A \otimes_K L \simeq M_n(L) \sim L$ *), then* L *is called a "splitting field of* A *(or of* $[A]$ *)".*

Theorem 3 implies

(3) $\qquad Br(L/K) \subseteq Br(M/K)$ if $K \subseteq L \subseteq M$, i.e. extensions of splitting fields of A are likewise splitting fields of A.

Moreover, Theorem 1(e) implies:

(4) $\qquad$ every central simple K-algebra has a splitting field.

Note that by definition the sequence of Z-modules

(5) $\qquad 0 \longrightarrow Br(L/K) \hookrightarrow Br(K) \xrightarrow{r_{L/K}} Br(L)$ is exact.

Now (1),...,(8) in §8. show

Lemma 4. *Let* L/K *be finite Galois and* $\Gamma := Gal(L/K)$ *, then the assignment* $^\sigma[A] := [^\sigma A]$ *(* $\sigma \in \Gamma$ *,* $[A] \in Br(L)$ *) endows* $Br(L)$ *with the structure of a left* Γ*-module such that* $\operatorname{Im} r_{L/K} \subseteq Br(L)^\Gamma$. □

Therefore we can amplify (5) and get

(6) $\qquad$ Let L/K be finite Galois with $\Gamma := Gal(L/K)$, then
$\qquad 0 \longrightarrow Br(L/K) \hookrightarrow Br(K) \xrightarrow{r_{L/K}} Br(L)^\Gamma$ is exact.

Using Theorem 7 in §7. in connection with Lemma 2 in §8. we conclude from

Lemma 4 above:

Theorem 4. *Let* L/K *be a finite cyclic extension with Galois group* Γ, *then the sequence*
$$0 \longrightarrow Br(L/K) \underset{=}{\hookrightarrow} Br(K) \xrightarrow{r_{L/K}} Br(L)^{\Gamma} \longrightarrow 0$$
is exact. □

Theorem 4 is due to O. Teichmüller [1940]; in the same paper the image of $r_{L/K}$ in the sequence (6) is also studied. The latter has been improved by S. Eilenberg & S. MacLane [1948] who extend the sequence (6) to the right by two terms (of course this is related to the Hochschild-Serre spectral sequence, cf. §18.) . Moreover, in §15. we shall discuss the sequence (5) extensively in the case " L/K purely inseparable" .

Theorem 3 has a counterpart in the finite separable case.

Theorem 5. *Let* L/K *be a finite separable extension, then the assignment* A ↦ $c_{L/K}$(A) *(according to Definition 2 in §8.) induces a* Z*-homomorphism* $c_{L/K}$: Br(L) ⟶ Br(K) *which is functorial in* L *(i.e.* $c_{M/L}c_{L/K}$ = = $c_{M/K}$ *if* K ⊆ L ⊆ M *) and such that*
$$c_{L/K}r_{L/K} = |L:K|id_{Br(K)}$$
and
$$r_{L/K}c_{L/K} = N_{\Gamma} \text{ if } L/K \text{ is Galois with } \Gamma := Gal(L/K) .$$

Proof. Clear from the various results in §8. □

Now let us restate *(v)* in the Centralizer Theorem in §7. in our new language:

Theorem 6. *Let* [A] ∈ Br(K) , [B] ∈ Br(L) *and* B ⊆ A *(L/K a finite field extension); then we have in* Br(L) *the equations*
$$[Z_A(B)] = [A \otimes_K L] - [B] = r_{L/K}([A]) - [B] . \square$$

Now consider in Theorem 6 the case "B = L" ; this gives

Corollary 1. *Let* L/K *be a finite extension and* [A] ∈ Br(K) *such that* L ⊆ A , *then* $[Z_A(L)] = [A \otimes_K L] = r_{L/K}([A])$. □

The following shows that the condition "B ⊆ A" in Theorem 6 is not a serious thing.

Lemma 5. *Let* L/K *be a finite extension,* [A] ∈ Br(K) *and* [B] ∈ Br(L) *then there is an* A' *such that* [A'] = [A] *and* B ⊆ A' .

Proof. Denote by L_b the left multiplication (b ∈ B) and consider the

K-algebra homomorphism (here $n := |A:K|$)

$$f: B \to A' := \text{End}_K(B) \otimes_K A \simeq M_n(K) \otimes_K A \sim A, \quad b \mapsto L_b \otimes 1.$$

Here f is injective (since B has no proper two-sided ideals), hence we can view B as embedded in A'. □

Now we turn our attention to questions of separability.

Lemma 6. *Assume $p := \text{char}(K) \neq 0$ and consider a K-skew field D; let $d \in D$ be such that $d \notin K$ but $d^p \in K$ (i.e. $K(d)/K$ is purely inseparable of degree p), then there exists a separable field extension L/K such that $K \neq L \subseteq D$.*

Proof. The assignment $x \mapsto dxd^{-1}$ defines an (inner) automorphism σ; by construction $\sigma \neq \text{id}_D$ but $\sigma^p = \text{id}_D$. Consider the endomorphism $\tau :=$
$:= \sigma - \text{id}_D$ of D: obviously $\tau \neq 0$ but $\tau^p = (\sigma - \text{id}_D)^p = \sigma^p - \text{id}_D = 0$
for reasons of characteristic. Set $r := \max\{i \mid \tau^i \neq 0\}$, hence $1 \leq r < p$; by construction there exists some $y \in D$ such that $\tau^r(y) \neq$
$\neq 0$. Set $a := \tau^{r-1}(y) \neq 0$ and $b := \tau(a) \neq 0$; it follows $\sigma(a) =$
$= \tau(a) + a = b + a$ and $\tau(b) = \tau^{r+1}(y) = 0$, hence $\sigma(b) = b$, and therefore - if $c := b^{-1}a \in D$ -: $\sigma(c) = \sigma(b^{-1})\sigma(a) = b^{-1}(b + a) = 1 + c$. Now consider the field $M := K(c) = K(1+c) \subseteq D$; M/K cannot be purely inseparable because it admits the K-algebra automorphism $\sigma|_M \neq \text{id}_M$, hence the separable closure L of K in M has the desired properties. □

Now let us draw two important conclusions from Lemma 6 and its proof; the first one concerns cyclic extensions of degree p of fields of characteristic p (we shall make use of this result later in §15.):

Artin-Schreier Theorem. *Let $p := \text{char}(K) \neq 0$ and consider a finite cyclic extension N/K of degree p with $\Gamma := \text{Gal}(N/K) = \sigma$; then $N = K(c)$ where c is such that $^\sigma c = 1+c$ and $a := c^p - c \in K^*$, i.e. c has the minimal polynomial $T^p - T - a \in K[T]$.*

Proof. Copy the proof of Lemma 6 with the following alteration: replace D by N and forget about the d (note: there we needed the d for the definition of σ, here the σ is given !). Finally conclude $L = M = N$ for reasons of degree and observe $a \neq 0$ as well as (for reasons of characteristic) $^\sigma a = (1+c)^p - (1+c) = c^p - c = a$, hence $a \in \text{Fix}_L(\sigma) =$
$= K$. □

The second application of Lemma 6 is

Corollary 2. *Let $D \neq K$ be a K-skew field, then there exists a separable field extension L/K such that $K \neq L \subseteq D$.*

Proof. In case $char(K) = 0$ there is nothing to prove. Now consider the case $p := char(K) \neq 0$; if there were no such extension L/K every $z \in D$ would have to be purely inseparable over K, hence there would be some $e \in \mathbb{N}$ (depending on z) such that $d := z^{p^e} \notin K$ but $d^p \in K$; this would contradict Lemma 6. □

From Corollary 2 we deduce the following important result.

Köthe's Theorem. *Let D be a K-skew field, then there exists a maximal commutative subfield $M \subseteq D$ such that M/K is separable.*

Proof. We proceed by induction on the index $i(D)$, the case "$i(D) = 1$" being trivial. Consider the case "$i(D) > 1$": take L as in Corollary 2 and consider $E := Z_D(L)$; E is an L-skew field (cf. Lemma 3 in §7.) of index $i(E) = i(D)/|L:K| < i(D)$ (cf. the Centralizer Theorem in §7.). By induction hypothesis E contains a maximal commutative subfield M such that M/L is separable, hence M/K is separable. Thanks to Theorem 4 in §7. M is also a maximal commutative subfield of D. □

In order to make full use of Köthe's Theorem we need a characterization of splitting fields of central simple algebras.

Theorem 7. *Let L/K be a finite extension and $[A] \in Br(K)$; then $[A] \in Br(L/K)$ (i.e. L is a splitting field of A) if and only if there exists an A' such that $[A'] = [A]$, $L \subseteq A'$ and $|L:K|^2 = |A':K|$.*

Proof. The "if" part is easy to prove: for reasons of degree we have $Z_{A'}(L) = L$ (cf. (10) in §7. and the Centralizer Theorem *ibid.*), hence
$$r_{L/K}([A]) = r_{L/K}([A']) = [L] = 0$$ thanks to Corollary 1.

For the "only if" part set $n^2 = |A:K| = |A\otimes_K L:L|$ and $m = |L:K|$; by assumption we have $A^{op} \otimes_K L \simeq M_n(L)$ (cf. Theorem 2). Now consider the *injective* K-algebra homomorphisms f,g (note that L and A^{op} have no proper two-sided ideals!) defined in the diagram below:

$$\begin{array}{c} x \\ \downarrow \\ 1\otimes x \\ a\otimes 1 \\ \uparrow \\ a \end{array} \quad \begin{array}{c} L \\ \downarrow \\ A^{op}\otimes_K L \simeq M_n(L) \simeq End_L(L^n) \subseteq End_K(L^n) \simeq M_{mn}(K) =: B \\ \uparrow \\ A^{op} \end{array} \xrightarrow{f} \xrightarrow{g}$$

Embed $L, A^{op} \subseteq B$ (see above) and define $A' := Z_B(A^{op})$ in this sense.

By construction it follows $L \subseteq A'$; Theorems 2/6 yield $[A'] = [B] - [A^{op}] = 0 - (-[A]) = [A]$ and finally *(vii)* in the Centralizer Theorem in §7. gives $|A':K| = m^2n^2/n^2 = m^2 = |L:K|^2$. □

If we combine Theorem 7 with Lemma 3(β) we get

Corollary 3. *Let* $[A] \in Br(L/K)$ *such that* $|L:K|^2 = |A:K|$, *then* $L \hookrightarrow A$, *i.e.* L *may be viewed as embedded in* A. □

If we apply Theorem 7 to the skew field component of A and use then Theorem 1*(f)* we get

Corollary 4. *Let* $[A] \in Br(K)$ *and* L *a splitting field of* A *of finite degree* $|L:K|$, *then* $i(A)$ *divides* $|L:K|$. □

Moreover, Theorem 7 and Theorem 4 in §7. imply

Corollary 5. *Any maximal commutative subfield* L *of a* K*-skew field* D *is a splitting field of* D. □

If we apply Corollary 5 to the skew field component of a central simple K-algebra A and make also use of Köthe's Theorem we get

Corollary 6. *Let* $[A] \in Br(K)$, *then there exists a splitting field* L *of* A *such that* L/K *is separable and* $|L:K| = i(A)$. □

Now the following amplification of Theorem 1 is obvious (cf. also (3)).

Theorem 8. *The following statements are equivalent:*
(a)   A *is a central simple* K*-algebra ;*
(g) $A \otimes_K L \simeq M_n(L)$ *for some finite separable* L/K ;
(h) $A \otimes_K N \simeq M_n(N)$ *for some finite Galois* N/K. □

Theorem 8*(h)* may be restated as

Theorem 9. $Br(K) = \bigcup_{\substack{L/K \text{ finite} \\ \text{Galois}}} Br(L/K)$. □

Here we remark that Theorems 8/9 can be strengthened in the sense that *Galois* can be replaced by the stronger *metabelian* (and hence *soluble*) - i.e. Galois with metabelian (and hence soluble) Galois group; cf. §§11/17.

We come to an important application of Corollary 6 (which depends on Köthe's Theorem):

Theorem 10. *Let* $[A] \in Br(K)$, *then* $i(A)[A] = 0$, *i.e.* $Br(K)$ *is a torsion group.*

Proof. Choose a splitting field L of A such that L/K is finite separable of degree i(A) (see Corollary 6); then Theorem 5 yields
$$i(A)[A] = |L:K|[A] = c_{L/K}r_{L/K}([A]) = c_{L/K}(0) = 0 \;.\;\square$$
Thanks to the preceding theorem the following definition is feasible:

Definition 5. *Let* [A] ∈ Br(K), *then* $o(A) := \min\{\, r \in \mathbb{N} \mid r[A] = 0 \,\}$ – *i.e. the order (in the sense of Group Theory) of* [A] *in* Br(K) – *is called the "exponent of* A *"*.

Of course we have (compare with (2))

(7) $o(A) = o(B)$ if $[A] = [B]$.

Moreover, Theorem 10 implies

Corollary 7. $o(A) \mid i(A)$ *for all* [A] ∈ Br(K). $\square$

Another easy consequence is (cf. Group Theory):

(8) $o(A \otimes_K L) \mid o(A)$ for [A] ∈ Br(K) (L/K arbitrary) and
$o(c_{L/K}(B)) \mid o(B)$ for [B] ∈ Br(L) (L/K finite separable).

Our next result may be viewed as a weakened converse to Corollary 7.

Theorem 11. *Let* [A] ∈ Br(K) *and let* p *be a prime dividing* i(A); *then* p *divides* o(A), *i.e. index and exponent have precisely the same prime factors (apart from multiplicities)*.

Proof. By Theorem 9 we may assume [A] ∈ Br(L/K) for some finite Galois L/K. Let $\Gamma := \mathrm{Gal}(L/K)$ and choose any p-Sylow subgroup Γ_p of Γ; let $L_p := \mathrm{Fix}_L(\Gamma_p)$ be the corresponding p-Sylow subfield of L, then
$$p \nmid |L_p:K| \text{ but } p \mid i(A) \mid |L:K| = |\Gamma| \text{ (cf. Corollary 4)},$$
hence
$$p^f = |\Gamma_p| = |L:L_p| > 1 \;.$$
Now $[A \otimes_K L_p] \in \mathrm{Br}(L/L_p)$ (cf. Theorem 3) and $0 \neq [A \otimes_K L_p]$ in $\mathrm{Br}(L_p)$ (for otherwise Corollary 4 would imply $p \mid i(A) \mid |L_p:K|$ which contradicts our construction), hence (thanks to Corollaries 4/7)
$$1 \neq o(A \otimes_K L_p) \mid i(A \otimes_K L_p) \mid |L:L_p| = p^f > 1 \;,$$
therefore $p \mid o(A \otimes_K L_p) \mid o(A)$ because of (8). $\square$

Incidentally, it was R. Brauer himself who showed first that no more relations between index and exponent than the ones coming from Corollary 7/ Theorem 11 can be established *in general*; cf. also (7) in §24.

We close this paragraph with some remarks on *index reduction*.

Theorem 12. *Let* $[A] \in Br(K)$ *and* L/K *a (not necessarily finite) field extension; then*

(A) $\quad i(A \otimes_K L) \mid i(A)$;

(B) $\quad \dfrac{i(A)}{i(A \otimes_K L)} \Big| |L:K|$ *in case* L/K *is a finite extension ;*

(C) $\quad \dfrac{i(A)}{i(A \otimes_K L)} = |L:K|$ *if and only if* L *can be embedded into the skew field component* D *of* A .

Definition 6. *The quotient* $\dfrac{i(A)}{i(A \otimes_K L)}$ *in Theorem 12 is called the "index reduction factor (of* A *relative to* L/K *)".*

Proof. Clearly we may restrict our attention to the case where $A = D$ is a skew field (cf. (2)). Now write $D \otimes_K L \simeq M_r(E)$ where the L-skew field E is the skew field component of the left hand side (cf. Theorem 1(f)) ; by definition $i(D \otimes_K L) = i(E)$, hence (A) and $r = \dfrac{i(D)}{i(D \otimes_K L)}$. Now set $m :=$
$:= |L:K|$ and consider the K-algebra homomorphism $L: L \to End_K(L)$, $x \mapsto L_x$ where L_x stands for left multiplication with x ; $id_D \otimes L$ is then an injection $M_r(E) \simeq D \otimes_K L \to D \otimes_K End_K(L) \simeq D \otimes_K M_m(K) =: B$, hence we may (and shall) view both L and $M_r(K)$ embedded in B . Now define $C :=$
$:= Z_B(M_r(K))$; it follows $L \subseteq C$ and (see Theorem 6)
$$[C] = [Z_B(M_r(K))] = [B] - [M_r(K)] = [B] = [D] ,$$
hence
$$L \subseteq C \simeq M_t(D) \quad \text{for suitable } t$$
thanks to Theorem 1(f). On the other hand we find $B \simeq C \otimes_K M_r(K)$ (cf. Corollary 8 in §7.), consequently (see also Lemma 4/Corollary 1 in §5.)
$$M_m(D) \simeq D \otimes_K M_m(K) = B \simeq C \otimes_K M_r(K) \simeq M_t(D) \otimes_K M_r(K) \simeq M_{rt}(D) ,$$
i.e. $|L:K| = m = rt$. This proves (B) and the "only if" part of (C) simultaneously. Finally, for the proof of the "if" in (C), we consider $E' :=$
$:= Z_D(L)$; then E' is an L-skew field such that $[E'] = [D \otimes_K L]$ satisfying $i(D) = i(E')m = i(D \otimes_K L)|L:K|$ (cf. the Centralizer Theorem/Lemma 3 in §7.). □

Corollary 8. *Let* $[A] \in Br(K)$ *and* L/K *a finite extension such that* $i(A)$ *and* $|L:K|$ *are coprime; then* $i(A \otimes_K L) = i(A)$. *In particular: if* D *is a K-skew field and* L/K *a finite extension such that* $|D:K|$ *and* $|L:K|$ *are coprime, then* $D \otimes_K L$ *remains a skew field.* □

Of course, Corollary 8 remains true for *infinite* extensions L/K if one interprets $|L:K|$ in this case as a *supernatural number* (see §2 in Ch.I

S.S. Shatz [1972]).

Corollary 9 . *If D is a K-skew field and L/K a finite extension of prime degree, then either $L \hookrightarrow D$ or $D \otimes_K L$ remains a skew field.* □

Now we investigate another type of index reduction:

Theorem 13 . *Let $[A],[B] \in Br(K)$, then*

(A) $i(A \otimes_K B) | i(A)i(B)$;

(B) $\frac{i(A)i(B)}{i(A \otimes_K B)} | i(A)^2, i(B)^2$.

Proof. Again we may restrict our attention to the case where A and B are K-skew fields (cf. (2)). Then we have (cf. Theorems 1/2)

$$A \otimes_K B \simeq D \otimes_K M_s(K) \quad \text{and} \quad A^{op} \otimes_K D \simeq B \otimes_K M_t(K)$$

for some K-skew field D and suitable $s, t \in \mathbb{N}$. Since $i(A \otimes_K B) = i(D)$ and $i(A)i(B) = i(D)s$ we get *(A)* immediately. As for *(B)* it suffices to prove (say) "$s | i(A)^2$" ; using Corollary 2 in §5. we obtain (here $n := i(A)^2 = |A:K|$)

$$M_n(B) \simeq M_n(K) \otimes_K B \simeq (A^{op} \otimes_K A) \otimes_K B \simeq A^{op} \otimes_K (A \otimes_K B) \simeq$$

$$\simeq A^{op} \otimes_K (D \otimes_K M_s(K)) \simeq (A^{op} \otimes_K D) \otimes_K M_s(K) \simeq B \otimes_K M_{st}(K) ,$$

hence $i(A)^2 = n = st$ (cf. the various rules from §§4/5.). □

Corollary 10 . *Let $[A],[B] \in Br(K)$ such that $i(A)$ and $i(B)$ are coprime, then $i(A \otimes_K B) = i(A)i(B)$. In particular: if D, E are K-skew fields of coprime index, then $D \otimes_K E$ remains a skew field.* □

Theorem 14 . *Let D be a K-skew field of index $i(D) = mn$ with coprime m and n ; then there exist unique (up to isomorphism) K-skew fields E and F such that $i(E) = m$, $i(F) = n$ and $E \otimes_K F \simeq D$.*

Proof. Using Corollary 7 we may write

$$o(D) = m_0 n_0 \quad \text{with coprime } m_0 \text{ and } n_0 \text{ such that } m_0 | m , n_0 | n$$

and find unique $[E],[F] \in Br(K)$ such that (compare the elementary theory of abelian torsion groups)

$$[D] = [E] + [F] \quad \text{where} \quad o(E) = m_0 \quad \text{and} \quad o(F) = n_0 .$$

Now $o(E)$ and $o(F)$ are coprime by our construction; consequently (cf. Theorem 11) $i(E)$ and $i(F)$ are likewise coprime, hence $E \otimes_K F$ is also a skew field $\sim D$. Now Lemma 3(δ) implies even $E \otimes_K F \simeq D$. □

Corollary 11 . *Let D be a K-skew field and $i(D) = \prod_{\rho=1}^{r} p_\rho^{f_\rho}$ the prime power factorization if its index, then*

$$D \simeq \bigotimes_{\rho=1}^{r} D_\rho \quad \textit{with K-skew fields} \quad D_\rho \quad \textit{such that} \quad i(D_\rho) = p_\rho^{f_\rho} \, . \quad \square$$

Now use (8) with $B := A \otimes_K L$ and apply Theorem 5 ; this gives

Corollary 12. *Let* $[A] \in Br(K)$ *and* L/K *a finite separable extension such that* $o(A)$ *and* $|L:K|$ *are coprime; then* $o(A \otimes_K L) = o(A)$. $\square$

Note that $o(A)$ and $|L:K|$ are coprime if and only if $i(A)$ and $|L:K|$ are (cf. Theorem 11).

Theorem 5 and Corollaries 8/12 can be restated in a more coherent way: denote by ${}_m Br(K) := \{ [A] \in Br(K) \mid m[A] = 0 \}$ the m-*torsion component* of $Br(K)$ - hence $[A] \in {}_m Br(K)$ if and only if $o(A) | m$ - and by $Br(K)_p :=$
$$:= \bigcup_{f=1}^{\infty} {}_{p^f} Br(K)$$
its p-*primary component* ($m \in \mathbb{N}$, p a prime), then

Corollary 13. *Let* L/K *be finite separable such that* m *and* $|L:K|$ *are coprime* ($m \in \mathbb{N}$), *then* $r_{L/K}: {}_m Br(K) \to {}_m Br(L)$ *is injective and preserves index as well as exponent. Moreover,* $c_{L/K}: {}_m Br(L) \to {}_m Br(K)$ *is surjective.* $\square$

Another way of expressing the same ideas is

Corollary 14. *Let* L/K *be finite Galois,* $\Gamma := Gal(L/K)$, Γ_p *a p-Sylow subgroup of* Γ *and* $L_p := Fix_L(\Gamma_p)$ *the corresponding p-Sylow subfield of* L , *then* $r_{L_p/K}: Br(L/K)_p \to Br(L/L_p)$ *is injective and preserves index as well as exponent. Moreover,* $c_{L_p/K}: Br(L/L_p) \to Br(L/K)_p$ *is surjective.* $\square$

Again the *finiteness* of L/K (in Corollaries 12/13/14) is not really necessary; all these results can easily be extended to the infinite case by inductive (resp. projective) limit techniques (cf. Ch.I of S.S. Shatz [1972].). The *separability* conditions are also superfluous; for instance Corollary 12 remains correct even in the inseparable case (cf. S.A. Amitsur [1962]).

Exercise 1. Prove the converse of the Artin-Schreier Theorem, namely: let $p := char(K) \neq 0$, $a \in K$ and consider the polynomial $f(T) :=$
$:= T^p - T - a \in K[T]$; then *either* $f(T)$ splits into p factors over K or $f(T)$ is irreducible and any root c generates a field $K(\frac{1}{p}a) := K(c)$ such that $K(\frac{1}{p}a)/K$ is cyclic with generating automorphism $\sigma: c \mapsto c + 1$.

Exercise 2. Let $[A] \in Br(K)$ and L/K a field extension, then

$$\frac{i(A)}{i(A\otimes_K L)} = r|L:K|$$ if and only if r is minimal such that L can be embedded in $M_r(D)$ where D stands for the skew field component of A.

(cf. Theorem 23 in Ch.IV of A.A. Albert [1939]; note that this exercise generalizes Theorem 12*(C)*)

§ 10 . CYCLIC ALGEBRAS

In §7. (cf. Definition 4 *ibid.*) we have already introduced the *cyclic algebras*

(1)
$$(a,L/K,\sigma) := \bigoplus_{i=1}^{n-1} Le^i \text{ with multiplication } e^n = a \text{ and}$$
$$ex = {}^{\sigma}xe \text{ (hence } xe^{-1} = e^{-1\sigma}x \text{) for all } x \in L \text{ ;}$$

here L/K is a finite cyclic extension of degree n with generating automorphism σ and a $\in$ K* is given. Thanks to Lemma 6 in §7. we have (with the notation introduced in §9.)

(2) $\qquad$ [a,L/K,σ] := [(a,L/K,σ)] $\in$ Br(L/K) $\subseteq$ Br(K) .

Now we claim (situation as above)

Lemma 1 . (a,L/K,σ) $\simeq$ (b,L/K,σ) *provided* $\frac{b}{a} \in N_{L/K}(L^*)$.

Proof. Write A := (a,L/K,σ) = $\bigoplus_i Le^i$ and B := (b,L/K,σ) = $\bigoplus_i Lf^i$; now take c $\in$ L* such that b = $aN_{L/K}(c)$ and set f_0 := ec $\in$ A* . It follows $f_0^n = (ec)^n = .. = N_{L/K}(c)a = b$ and $f_0 x = ecx = {}^{\sigma}xec = {}^{\sigma}xf_0$ for all x $\in$ L , hence the assignment f $\mapsto f_0$, x $\mapsto$ x (x $\in$ L) induces a K-algebra homomorphism g: B $\to$ A which is even an isomorphism because of Lemma 1 in §9. □

Lemma 2 . *Let* [A] $\in$ Br(K) *and* L/K *a cyclic extension of degree* n *with generating automorphism* σ *such that* L $\subseteq$ A *and* $n^2 = |A:K|$; *then there exists an* e $\in$ A* *such that* a := $e^n \in K^*$, ex = ${}^{\sigma}$xe (x $\in$ L) *and* A = $\bigoplus_{i=0}^{n-1} Le^i$. *Moreover, if* f $\in$ A* *is such that* b := $e^n \in K^*$, fx = ${}^{\sigma}$xf (x $\in$ L) *and* A = $\bigoplus_{i=0}^{n-1} Lf^i$, *then* $\frac{b}{a} \in N_{L/K}(L^*)$.

Proof. By Theorem 3 in §7. we see L = $Z_A(L)$, hence the existence of e $\in$ A* such that A = $\bigoplus_i Le^i$ is already clear from Theorems 5/6 in §7. Inspection of the proof of Theorem 5 *ibid.* shows that e arises from the

Skolem-Noether Theorem in §7.:
$$^\sigma x = exe^{-1} \quad \text{for all} \quad x \in L .$$
It follows $x = {^{\sigma^n}}x = e^n x e^{-n}$ for all $x \in L$, hence $a := e^n \in Z_A(L) = L$
and therefore $^\sigma a = ee^n e^{-1} = e^n = a$, hence $a \in \text{Fix}_L(\sigma) = K$. Now if
$$fxf^{-1} = {^\sigma x} = exe^{-1} \quad \text{for all} \quad x \in L ,$$
then $e^{-1}fxf^{-1}e$ and consequently $c := e^{-1}f \in Z_A(L) = L$ as well as
$$b := f^n = (ec)^n = \ldots = N_{L/K}(c)e^n = N_{L/K}(c)a . \quad \square$$

If we replace in the preceding proof σ by another generator σ^t (note that t and n must be coprime) of $\text{Gal}(L/K)$, then e must be replaced by e^t, hence a by a^t. This gives

Lemma 3. *For cyclic algebras we have the rule*
$$(a,L/K,\sigma) \simeq (a^t,L/K,\sigma^t) \quad \text{provided } t \text{ and } |L:K| \text{ are coprime.} \quad \square$$

The next rule is crucial.

Lemma 4. *For cyclic algebras we have the rule* ($n = |L:K|$)
$$(a,L/K,\sigma) \otimes_K (b,L/K,\sigma) \simeq M_n\big((ab,L/K,\sigma)\big) .$$

Proof. Set $A := (a,L/K,\sigma) = \bigoplus_i Le^i$, $B := (b,L/K,\sigma) = \bigoplus_i Lf^i$ and
$C := (ab,L/K,\sigma) = \bigoplus_i Le_0^i$; now consider the matrices

(3) $\quad A_j := \begin{pmatrix} 0 & 1 & 0 \\ & 0 & 1 \\ \hline a & 0 & \\ 0 & a & 0 \end{pmatrix} \in M_n(K)$ for $j = 0,\ldots,n$, define $A := A_1$ and observe $A_j = A^j$ (j as above).

$\longleftarrow$ n-j rows and columns, i-th row and column
$\longleftarrow$ j rows and columns

If we define $g(e) := A$, $h(f) := e_0 A^{-1} = A^{-1} e_0$, $g(x) := \begin{pmatrix} \ddots & & 0 \\ & {^{\sigma^i}}x & \\ 0 & & \ddots \end{pmatrix}$ and
$h(y) := y1$ ($x,y \in L$),
then these matrices in $M_n(C)$ satisfy the following equations:
$g(e)^n = A^n = a$ (cf. (3)), $g(e)g(x) = g({^\sigma x})g(e)$ (cf. Lemma 1 in §2.), $h(f)^n = A^{-n}e_0^n = a^{-1}ab = b$ and $h(f)h(y) = A^{-1}e_0 y = {^\sigma y}A^{-1}e_0 = h({^\sigma y})h(f)$,
hence two K-algebra homomorphisms $g: A \to M_n(C)$, $h: B \to M_n(C)$ are well-defined. Now our claim is an easy consequence of Lemma 2 in §9. thanks to
$g(e)h(f) = e_0 = h(f)g(e)$, $g(x)h(y) = h(y)g(x)$, $g(e)h(y) = h(y)g(e)$ and $h(f)g(x) = g(e)^{-1}e_0 g(x) = g(e)^{-1}g({^\sigma x})e_0 =$

$$= g(x)g(e)^{-1}e_0 = g(x)h(f) \quad (x,y \in L) . \quad \square$$

Theorem 1. *Let* L/K *be finite cyclic with generating automorphism* σ, *then the assignment* $a \mapsto [a,L/K,\sigma]$ ($a \in K^*$, cf. (2)) *induces an isomorphism*

$$\Theta_\sigma : K^*/N_{L/K}(L^*) \xrightarrow{\sim} Br(L/K) \quad .$$

Note that in view of Example 1 in §6. the latter isomorphism may be restated in the form

(4) $\quad H^0(\Gamma, L^*) \simeq Br(L/K) \quad$ in case $\quad \Gamma := Gal(L/K)$ is cyclic.

Proof. Thanks to (2) and Lemma 1 the map Θ_σ is well-defined. Lemma 4 says that Θ_σ is even a homomorphism. The injectivity of Θ_σ follows from Lemma 2 whereas the surjectivity is a consequence of Lemma 2 and Theorem 7 in §9. $\square$

Now we are ready for the proof of two *really classical* results:

Frobenius' Theorem (1878). *Let* D *be an R-skew field* $\neq R$, *then necessarily* $D \simeq \mathbb{H}$ (cf. (1) in §1.), *i.e.* $Br(R) \simeq Z/2Z$.

Proof. We have $Br(R) = Br(C/R) \simeq R^*/N_{C/R}(C^*) \simeq R^*/R^*_{>0} \simeq Z/2Z$ (see Theorem 9 in §9. and Theorem 1 above), hence there is just one R-skew field $\neq R$ which must (up to isomorphism) be the one introduced in §1. $\square$

Wedderburn's Theorem (1905). *All finite skew fields are commutative, i.e.* $Br(K) = \{0\}$ *if* K *has only finitely many elements.*

Proof. We make use of the following result from Field Theory (which we shall prove below for the convenience of the reader):

Lemma 5. *Let* K *be a finite field and* L/K *a finite field extension; then* L/K *is necessarily cyclic with surjective norm* $N_{L/K} : L^* \to K^*$.

Proof. Set $q := |K|$, $n := |L:K|$, hence $q^n = |L|$. It is well-known that $x \mapsto x^q$ defines a K-automorphism of L - called the "Frobenius automorphism" - which is clearly of order n; therefore L/K must be cyclic (cf. Artin's Lemma in §6.) with the above automorphism as a generator of its Galois group. It follows

$$N_{L/K}(x) = \prod_{i=0}^{n-1} x^{q^i} = x^{\frac{q^n-1}{q-1}} = x^{|L^*:K^*|}$$

where $|L^*:K^*|$ denotes the index of the multiplicative group K^* in L^*. Since L^* is known to be a cyclic group (cf. Field Theory) $N_{L/K}$ must be surjective. $\square$

Proof of Wedderburn's Theorem (continued). Thanks to Theorem 9 in §9., Lemma 5 and Theorem 1 we obtain

$$Br(K) = \bigcup_{\substack{L/K \text{ finite} \\ \text{Galois}}} Br(L/K) = \bigcup_{\substack{L/K \text{ finite} \\ \text{cyclic}}} Br(L/K) \simeq$$

$$\simeq \bigcup_{\substack{L/K \text{ finite} \\ \text{cyclic}}} K^*/N_{L/K}(L^*) \simeq \{0\} \quad . \quad \square$$

Of course, J.H.M. Wedderburn's original proof was along different lines; note that there exists a simple *direct proof* (i.e. a proof which makes no use of the concept of Brauer groups) due to E. Witt ; it is reproduced e.g. in the first paragraph of A. Weil [1967].

Now let us deepen our study of the formalism of cyclic algebras.

Lemma 6 . *Let* L/K *be cyclic of degree* n *with generating automorphism* σ *and let* I *be an intermediate field of degree* m *over* K *; then* σ^m *generates* Gal(L/I) *and*

$$(a,L/K,\sigma) \otimes_K I \sim (a,L/I,\sigma^m) \ .$$

Proof. Write $A := (a,L/K,\sigma) = \bigoplus_i Le^i$ and consider $Z_A(I)$; now let $z = \sum_{i=0}^{n-1} x_i e^i \in A$ be given. Then $z \in Z_A(I)$ if and only if

$$0 = zx - xz = \sum_i (\sigma^i x - x)x_i e^i \quad \text{for all} \quad x \in I = \text{Fix}_L(\sigma^m) \ ,$$

hence if and only if $x_i = 0$ for all i which are not multiples of m . The latter amounts to ($s := n/m = |L:I|$)

$$z = \sum_{j=0}^{s-1} x_{mj} e^{mj} = \sum_{j=0}^{s-1} y_j f^j \quad (\ y_j := x_{mj} \ , \ f := e^m \ )$$

where $f^s = e^{ms} = a$ and $fy = e^m y = \sigma^m y e^m = \sigma^m y f$, hence (cf. also Corollary 1 in §9.)

$$(a,L/I,\sigma^m) \simeq Z_A(I) \sim A \otimes_K I = (a,L/K,\sigma) \otimes_K I \ . \quad \square$$

The next result is a supplement to the previous one.

Lemma 7 . *Let* L/K *be cyclic of degree* n *with generating automorphism* σ *and let* F/K *be an arbitrary (not necessarily finite) extension such that* L ∩ F = K *. Then* $L \otimes_K F$ *is a field* ≃ LF *such that* $L \otimes_K F/F$ *is likewise cyclic of degree* n *with generating automorphism* σ⊗id_F *(which we abbreviate by* σ *after identifying* $L \otimes_K F$ *with* LF *) and such that*

$$(a,L/K,\sigma) \otimes_K F \simeq (a,L\otimes_K F/F, \sigma \otimes id_F) = (a,LF/F,\sigma) \ .$$

Proof. All but the last assertion is well-known from Field Theory. As for the formula concerning the cyclic algebras we note that there is an obvious injection $(a,L/K,\sigma) \to (a,L\otimes_K F/F, \sigma \otimes id_F)$; F-linear extension thereof (cf. Theorem 3 in §5.) and Lemma 1 in §9. give the required isomorphism. □

Lemma 8. *Let L/K be cyclic of degree n with generating automorphism σ and let I be an intermediate field of degree m over K; then $\bar{\sigma} := \sigma|_I$ generates $\mathrm{Gal}(I/K)$ and*
$$(a^s, L/K, \sigma) \simeq M_s\big((a, I/K, \bar{\sigma})\big) \quad \text{where} \quad s := \frac{n}{m} = |L:I| \ .$$

Proof. Let $\{e_1,\ldots,e_s\}$ be a basis of L over I, then - if we write $e := (e_1,\ldots,e_s) \in L^s$ - every $x \in L$ determines a matrix
$$f(x) \in M_s(I) \quad \text{via} \quad xe = ef(x) \ .$$
On the other hand we have a map $F: \Gamma := \mathrm{Gal}(L/K) \to GL_s(I)$ via the correspondence $^\tau e = eF(\tau)$; it follows at ease ($\tau, \rho \in \Gamma$; $x \in L$)
$$ef(^\tau x)F(\tau) = {}^\tau xeF(\tau) = {}^\tau x\,{}^\tau e = {}^\tau(xe) = {}^\tau(ef(x)) = {}^\tau e\,{}^\tau f(x) =$$
$$= eF(\tau)\,{}^\tau f(x) \quad \text{and} \quad eF(\tau\rho) = {}^{\tau\rho}e = {}^\tau({}^\rho e) = {}^\tau(eF(\rho)) =$$
$$= {}^\tau e\,{}^\tau F(\rho) = eF(\tau)\,{}^\tau F(\rho)$$
(note that Γ acts on matrices componentwise), hence
(5) $\quad f(^\tau x)F(\tau) = F(\tau)\,{}^\tau f(x)$ for all $\tau \in \Gamma$, $x \in L$
and (note $\sigma^n = 1$)
(6) $\quad F(\tau\rho) = F(\tau)\,{}^\tau F(\rho)$, in particular $1 = F(\sigma)^\sigma F(\sigma)..^{\sigma^{n-1}}F(\sigma)$.
Now set $A := (a^s, L/K, \sigma) = \bigoplus_{i=1}^{n-1} Le^i$, $\bar{A} := (a, I/K, \bar{\sigma}) = \bigoplus_{j=1}^{m-1} I\bar{e}^j$ and define $f(e) := F(\sigma)\bar{e} \in M_s(\bar{A})$. Thanks to
$$f(e)f(x) = F(\sigma)\bar{e}f(x) = F(\sigma)^\sigma f(x)\bar{e} = f(^\sigma x)F(\sigma)\bar{e} = f(^\sigma x)f(e)$$
for all $x \in L$ (cf. (5)) and
$$f(e)^n = (F(\sigma)\bar{e})^n = \ldots = F(\sigma)^\sigma F(\sigma)..^{\sigma^{n-1}}F(\sigma)\bar{e}^n = \bar{e}^{ms} = a^s$$
(cf. (6)) we see that a K-algebra homomorphism $f: A \to M_s(\bar{A})$ is well-defined and even an isomorphism (because of Lemma 1 in §9.). □

We close this paragraph with a lemma which will be needed later in §15.:

Lemma 9. *Let L_j/K be cyclic of degree n_j with generating automorphism σ_j ($j = 1,2$) such that $L_1 \cap L_2 = K$ and $n_1 r = n_2$ ($r \in \mathbb{N}$); then $L := L_1 \otimes_K L_2$ is a field. Moreover, if $L_0 := \mathrm{Fix}_L(\sigma_1^{-1} \otimes \sigma_2^r)$ and $\sigma_0 := (id_{L_1} \otimes \sigma_2)|_{L_0}$, then L_0/K is cyclic with generator σ_0 and*
$$(a, L_1/K, \sigma_1) \otimes_K (a, L_2/K, \sigma_2) \simeq (1, L_1/K, \sigma_1^{-1}) \otimes_K (a, L_0/K, \sigma_0) \ .$$

Proof. The fact that L is a field is well-known from Galois Theory. Moreover, since $\sigma_1^{-1}\otimes\sigma_2^r$ has order n_1 we conclude $|L:L_0| = n_1$, hence $n_0 := |L_0:K| = n_2$. Now let $x_0 = \sum_\nu x_{1\nu}\otimes x_{2\nu} \in L$ ($x_{1\nu} \in L_1$, $x_{2\nu} \in L_2$) be given, then

$$x_0 \in L_0 \text{ if and only if } \sum_\nu \sigma_1^{-1}(x_{1\nu})\otimes\sigma_2^r(x_{2\nu}) = \sum_\nu x_{1\nu}\otimes x_{2\nu} ,$$

hence σ_0 has order $n_0 = n_2$, i.e. L_0/K is cyclic with generating automorphism σ_0.

Now write $A_j := (a,L_j/K,\sigma_j) = \bigoplus_i L_j e_j^i$ ($j = 0,1,2$), $A := (1,L_1/K,\sigma_1^{-1})$
$= \bigoplus_i L_1 e^i \simeq M_{n_1}(K)$ (cf. Theorem 1) and $B := A_1 \otimes_K A_2$; moreover, define embeddings

$$f: L_1 \hookrightarrow L \hookrightarrow B \text{ and } g: L_0 \hookrightarrow L \hookrightarrow B$$

as well as elements

$$f(e) := e_1^{-1}\otimes e_2^r \text{ and } g(e_0) := 1\otimes e_2 .$$

It follows ($x_1 \in L_1$, $x_0 \in L_0$) using the usual identification of $K \otimes_K K$ with K :

$$f(e)^{n_1} = a^{-1}\otimes a = 1 , \; g(e_0)^{n_0} = 1\otimes a = a , \; f(e)f(x_1) =$$
$$= e_1^{-1}x_1\otimes e_2^r = \sigma_1^{-1}(x_1)e_1^{-1}\otimes e_2^r = f(\sigma_1^{-1}(x_1))f(e) \text{ and } g(e_0)g(x_0) =$$
$$= \sum_\nu x_{1\nu}\otimes e_2 x_{2\nu} = \sum_\nu x_{1\nu}\otimes\sigma_2(x_{2\nu})e_2 = g(\sigma_0(x_0))g(e_0) .$$

Therefore the two embeddings f,g (see above) extend to K-algebra homomorphisms $f: A \to B$ and $g: A_0 \to B$. Now our claim follows from Lemma 2 in §9. thanks to

$$f(e)g(e_0) = g(e_0)f(e) , \; f(x_1)g(x_0) = g(x_0)f(x_1) , \; g(e_0)f(x_1) =$$
$$= f(x_1)g(e_0) \text{ (all three obvious) and } f(e)g(x_0) =$$
$$= \sum_\nu e_1^{-1}x_{1\nu}\otimes e_2^r x_{2\nu} = \sum_\nu \sigma_1^{-1}(x_{1\nu})e_1^{-1}\otimes\sigma_2^r(x_{2\nu})e_2^r =$$
$$= \sum_\nu x_{1\nu}e_1^{-1}\otimes x_{2\nu}e_2^r = g(x_0)f(e) . \quad \square$$

Exercise 1. Let L/K be cyclic of degree r with generating automorphism σ. Assume $r = mn$ where m and n are coprime. Now denote by M (resp. N) *the* intermediate field of L/K of degree m (resp. n) over K with generating automorphism $\tau := \sigma|_M$ (resp. $\rho := \sigma|_N$). Show

$$(a,L/K,\sigma) \simeq (b,M/K,\tau) \otimes_K (c,N/K,\rho)$$

for suitable $b,c \in K^*$ (depending on the given $a \in K^*$).

§ 11 . POWER NORM RESIDUE ALGEBRAS

Denote by μ_n the full group of n-th roots of unity and assume *throughout this paragraph* $\mu_n \subset K$ (note that this implies char(K)∤n). Now choose a,b $\in$ K* and select an auxiliary field extension L/K such that b $\in$ (L*)^n (for instance take L a separable closure of K). Furthermore denote by ζ a primitive n-th root of unity, i.e. $\langle\zeta\rangle = \mu_n$, and consider the matrices

$A \in M_n(K)$ according to (3) in §10., $Z := \begin{pmatrix} \ddots & & 0 \\ & \zeta^i & \\ 0 & & \ddots \end{pmatrix} \in M_n(K)$

and $B := БZ \in M_n(L)$ for some

Б $\in$ L such that $Б^n = b$.   i-th row and column

Lemma 1 in §2. yields

(1) $A^n = a$, $B^n = b$, and $AB = \zeta BA$

as well as the following obvious consequences thereof:

(2) $A^i B^j = \zeta^{ij} B^j A^i$, hence

$A^i B^j A^{-i} B^{-j} = \zeta^{ij} = (ABA^{-1}B^{-1})^{ij}$ ($i,j \in \mathbb{Z}$) .

Now consider the n^2 matrices $A^i B^j$ ($0 \leq i,j < n$); we want to show their L-linear independence (and therefore their K-linear independence): indeed, consider an L-linear combination

$$0 = \sum_{i=0}^{n-1}(\sum_{j=0}^{n-1} x_{ij} A^i B^j) \quad (x_{ij} \in L);$$

since the r-th row of $A^i = A_i$ (cf. (3) in §10.) has no entries $\neq 0$ outside the (i+r)-th column (read the indices *mod* n) the same statement remains true with

$$\sum_{j=0}^{n-1} x_{ij} A^i B^j \quad \text{in place of} \quad A^i \quad (i \in \{0,1,..,n-1\} \text{ fixed })$$

(cf. Lemma 1 in §2.), hence it suffices to prove the L-linear independence of the n elements $A^i B^j$ (j = 0,1,..,n-1) for every fixed i , i.e. - since $A \in GL_n(K) := M_n(K)^*$ - the L-linear independence of the n elements B^j (j = 0,1,..,n-1). The latter, however, amounts to the (well-known) non-vanishing of the Vandermonde determinant det(v_{ij}) where $v_{ij} =$

$= (b\zeta^i)^{j-1}$ ($1 \leq i,j \leq n$). Consequently

(3) $\bigoplus_{i=0}^{n-1} \bigoplus_{j=0}^{n-1} KA^i B^j =: A$ (note $A \subseteq M_n(L)$)

is an n^2-dimensional K-algebra such that L-linear extension of the embedding (cf. Theorem 3 in §5.) defines an L-algebra isomorphism $A \otimes_K L \xrightarrow{\sim}$ $\xrightarrow{\sim} M_n(L)$ (cf. Lemma 1 in §9.), hence A is even a central simple K-algebra (cf. Theorem 1(d) in §9.) which is completely determined (up to K-algebra isomorphism) by the formulae in (1). Since any K-algebra A generated by $a,b \in A$ and subject to the relations $a^n = a$, $b^n = b$ and $ab = \zeta ba$ is necessarily a homomorphic image of A (by (3)), it must be $\simeq A$ thanks to Lemma 1 in §9. Thus we have proved

Theorem 1 . *If* $\langle\zeta\rangle = \mu_n \subset K$, *then all K-algebras* A *generated by two elements* a,b *such that*

(4) $a := a^n \in K^*$, $b := b^n \in K^*$, $ab = \zeta ba$

are isomorphic (namely $\simeq A$ *in* (3)). *They are central simple K-algebras, and any field* L *such that* $K \subseteq L$ *and* $b \in (L^*)^n$ *is a splitting field.*□

Definition 1 . *We denote any of the isomorphic K-algebras described in Theorem 1 by*

$$(a,b;n,K,\zeta)$$

and call it (for reasons which will become clear later in this paragraph) a "power norm residue algebra". Moreover, we write

$$[a,b;n,K,\zeta] := [(a,b;n,K,\zeta)] \in Br(K) .$$

If no confusion can arise we omit any of the symbols n,K,ζ *in* $(a,b;n,K,\zeta)$ *and* $[a,b;n,K,\zeta]$.

Definition 2 . *If* char(K) $\neq$ 2 , $n = 2$, *hence* $\zeta = -1$, *we write*

$$\left(\frac{a,b}{K}\right) \text{ in place of } (a,b;2,K,-1)$$

and call such an algebra a "quaternion algebra".

See §14. for more details on quaternion algebras.

Lemma 1 . *Let* $\langle\zeta\rangle = \mu_n \subset K$, $|K(\sqrt[n]{a}):K| = n$ *and* $[A] \in Br(K(\sqrt[n]{a})/K)$, *then* $[A] = [a,b;n,K,\zeta]$ *in* $Br(K)$ *for suitable* $b \in K^*$.

Proof. In our situation it is known from Field Theory that $K(\sqrt[n]{a})/K$ is cyclic with generating automorphism $\sigma: \sqrt[n]{a} \mapsto \zeta\sqrt[n]{a}$. Now we replace A by B such that $[A] = [B]$, $a := \sqrt[n]{a} \in B$ and $|B:K| = n^2$ (cf. Theorem 7 in §9.). Then Lemma 2 in §10. implies the existence of $b \in B^*$ such that

$$B = \bigoplus_{j=0}^{n-1} K(a)b^j = \bigoplus_{j=0}^{n-1}(\bigoplus_{i=0}^{n-1} Ka^i)b^j \quad \text{where} \quad b^{-1} := b^n \in K^* \quad \text{and}$$

$bx = {}^\sigma x b$ for all $x \in K(a)$, in particular $ba = \zeta ab$, hence $ab^{-1} = \zeta b^{-1} a$ and therefore $B \simeq (a,b;n,K,\zeta)$ because of Theorem 1. □

Note that the above proof includes a proof of the following

Lemma 2. Let $\langle\zeta\rangle = \mu_n \in K$ and $|K(\sqrt[n]{a}):K| = n$, then
$$(a,b;n,K,\zeta) \simeq (b^{-1}, K(\sqrt[n]{a})/K, \sigma) \simeq (b, K(\sqrt[n]{a})/K, \sigma^{-1})$$
where ${}^\sigma(\sqrt[n]{a}) = \zeta\sqrt[n]{a}$. □

Moreover, since in case $\mu_n \subset K$ the cyclic extensions L/K of degree n coincide with the radical extensions (as in Lemma 1) - see e.g. Prop.5 on p.205 in vol.2 of P.M. Cohn [1974/77]; note that this is also an immediate consequence of Hilbert's "Satz 90" in §6. thanks to $N_{L/K}(\zeta) = 1$ -, we get from Lemma 1

Corollary 1. Let $\langle\zeta\rangle = \mu_n \subset K$, L/K *a cyclic extension of degree* n *and* $[A] \in Br(L/K)$, *then*
$$[A] = [a,b;n,K,\zeta] \text{ for suitable } a,b \in K^* . \quad □$$

Now we turn to a discussion of the main features of the power norm residue algebras.

Lemma 3. $(a,b) \otimes_K (a',b') \simeq (aa',b) \otimes_K (a',b^{-1}b')$, *i.e.* $[a,b] + [a',b'] = [aa',b] + [a',b^{-1}b']$ *in* $Br(K)$.

Proof. Write $A := (a,b) := (a,b;n,K,\zeta)$, $A' := (a',b') := (a',b';n,K,\zeta)$, and let a,b resp. a',b' be generators of A resp. A' (cf. Theorem 1). Now consider the elements

$$c := a \otimes a' , \quad d := b \otimes 1 , \quad c' := 1 \otimes a' \text{ and } d' := b^{-1} \otimes b'$$

in $A \otimes_K A'$. Clearly we obtain (by straightforward calculations)

(5) $\quad c^n = a \otimes a' = aa' , \quad d^n = b \otimes 1 = b , \quad cd = \zeta dc$

and

(6) $\quad c'^n = 1 \otimes a' = a' , \quad d'^n = b^{-1} \otimes b' = b^{-1}b' , \quad c'd' = \zeta d'c'$

(here we used the usual identification of $K \otimes_K K$ with K).
Moreover, all elements in (5) commute with all elements in (6). Hence the K-algebra homomorphisms $f: (aa',b) \to A \otimes_K A'$ and $g: (a',b^{-1}b') \to A \otimes_K A'$ (which exist thanks to (5),(6) and Theorem 1) define the required isomorphism because of Lemma 2 in §9. □

Now recall that

(7) $\quad [a,b] = 0$ if $b \in (K^*)^n$, i.e. n is a non-zero n-th power

(see Theorem 1: take $L = K$), set $b' = b$ in Lemma 3 and use (7). It follows immediately

(8) $\qquad [a,b] + [a',b] = [aa',b] + [a',1] = [aa',b]$.

Now take $a' = 1$ in (8); this gives

(9) $\qquad [a,b] + [1,b] = [a,b]$, hence $[1,b] = 0$.

Using Lemma 3 in the case $a' = a^{-1}$ and $b' = 1$ we find with the help of (7) and (9) the formula

(10) $\qquad [a,b] = [a,b] + [a^{-1},1] = [1,b] + [a^{-1},b^{-1}] = [a^{-1},b^{-1}]$

and consequently (use (9),(10) and apply Lemma 3)

(11) $\qquad [a,b] + [a,b'] = [a,b] + [a^{-1},b'^{-1}] = [1,b] + [a^{-1},b^{-1}b'^{-1}] =$
$\qquad = [a,bb']$.

Combining (7),(8) and (11) gives

Lemma 4. *Let* $\langle \zeta \rangle = \mu_n \subset K$, *then the assignment* $(a,b) \mapsto [a,b;n,K,\zeta]$ *induces a* $\mathbb{Z}$-*bilinear map* $K^*/(K^*)^n \times K^*/(K^*)^n \to {}_n\mathrm{Br}(K)$. □

So far the primitive n-th root of unity ζ was fixed. Now we want to study the impact of the change of ζ:

Lemma 5. *Let* n *and* t *be coprime, then* $(a,b;n,\zeta) \simeq (a,b^t;n,\zeta^t)$, *i.e.* $[a,b;n,\zeta] = t[a,b;n,\zeta^t]$ *in* $\mathrm{Br}(K)$.

Proof. Let a,b generators of $A := (a,b;\zeta)$ as in (4). Define
$$c := a \quad \text{and} \quad d := b^t \quad \text{in} \quad A .$$
Because of (2) this gives
$$c^n = a , \quad d^n = b^{tn} = b^t \quad \text{and} \quad cd = ab^t = \zeta^t b^t a = \zeta^t dc ,$$
hence $A \simeq (a,b^t;n,K,\zeta^t)$ by Theorem 1. □

An important consequence of Lemma 5 is

Corollary 2. *Let* $\mu_n \subset K$, *then the assignment* $(a,b) \mapsto [a,b;n,K,\zeta] \otimes \zeta$ *defines a bimultiplicative map* $K^* \times K^* \to {}_n\mathrm{Br}(K) \otimes_{\mathbb{Z}} \mu_n$ *which is independent of the choice of the primitive* n-*th root of unity* ζ. □

Moreover, since (by Theorem 1) obviously $(b,a;\zeta^{-1}) \simeq (a,b;\zeta)$, we obtain from Lemmata 4/5

Corollary 3. $[b,a] = -[a,b]$, *i.e.* $(b,a) \simeq (a,b)^{op}$. □

The next result is of utmost importance:

Lemma 6. *Assume* $n = rm$ *and* $\langle \zeta \rangle = \mu_n \subset K$, *hence* $\langle \zeta^r \rangle = \mu_m \subset K$, *then* $(a,b^r;n,K,\zeta) \simeq M_r\big((a,b;m,K,\zeta^r)\big)$, *i.e.* $r[a,b;n,K,\zeta] = [a,b;m,K,\zeta^r]$.

Proof. Let a,b be generators of $A := (a,b;m,\zeta^r)$ such that (cf. (4))

$$a^m = a, \quad b^m = b \quad \text{and} \quad ab = \zeta^r ba$$

and consider the matrices

$A \in M_r(A)$ according to (3) in §10. *but* with $a \in A$ in place of $a \in K$,

$B := bX \in M_r(A)$ where $X := \begin{pmatrix} \ddots & & 0 \\ & \zeta^i & \\ 0 & & \ddots \end{pmatrix} \in M_r(K)$ i-th row and column

(note that X coincides with the matrix Z at the beginning of this paragraph if $m = 1$). Clearly we have (after an easy calculation; see in particular Lemma 1 in §2.)

$$A^n = A^{rm} = a^m = a, \quad B^n = b^{mr} = b^r \quad \text{and} \quad AB = \zeta BA,$$

hence $M_r(A) \simeq (a, b^r; n, K, \zeta)$ because of Theorem 1. □

A consequence of Lemmas 5/6 is (cf. also Theorem 14 in §9.)

Lemma 7. *Let r, s be coprime, i.e. $1 = xs + yr$ for some $x, y \in \mathbb{Z}$, assume $\langle \xi \rangle = \mu_r \subset K$, $\langle \eta \rangle = \mu_s \subset K$, and set $n := rs$, $\zeta := \xi \eta$. Then $\langle \zeta \rangle = \mu_n \subset K$, $\xi = \zeta^{xs}$ and $\eta = \zeta^{yr}$. Moreover, we have*

$$(a, b; n, \zeta) \simeq (a^x, b^x; r, \xi) \otimes_K (a^y, b^y; s, \eta) \text{ which is equivalent to}$$
$$[a, b; n, \zeta] = x^2 [a, b; r, \xi] + y^2 [a, b; s, \eta] \text{ in } Br(K).$$

Proof. The fact that the last two assertions in the lemma are equivalent is clear from Lemma 3(β) in §9. Now Lemmas 5/6 imply

$$[a, b; n, \zeta] = xs[a, b; n, \zeta] + yr[a, b; n, \zeta] = x[a, b; r, \zeta^s] +$$
$$+ y[a, b; s, \zeta^r] = x^2[a, b; r, \xi] + y^2[a, b; s, \eta]. \quad □$$

Of course, Lemma 7 means that one may restrict the attention to the case " $n = p^t$ (p a prime $\neq$ char(K))".

Now, if L/K is an arbitrary (not necessarily finite) extension, we see that L-linear extension of the obvious embedding $(a, b; K) \hookrightarrow (a, b; L)$ yields the L-algebra isomorphism described in the lemma below (cf. Theorem 3 in §5. and Lemma 1 in §9.):

Lemma 8. $(a, b; K) \otimes_K L \simeq (a, b; L)$, *i.e.* $r_{L/K}[a, b; K] = [a, b; L]$. □

Combining Lemmas 4/6/8 gives then

Lemma 9. *Assume $n = rm$ and $\langle \zeta \rangle = \mu_n \subset K$, hence $\langle \zeta^r \rangle = \mu_m$; then*

$$r_{K(\sqrt[r]{a})/K}[a, b; n, K, \zeta] = [\sqrt[r]{a}, b; m, K(\sqrt[r]{a}), \zeta^r] \text{ in } Br(K(\sqrt[r]{a})). \quad □$$

Now consider Lemma 9 and use Galois Theory: the field $K(\sqrt[r]{a})$ is Galois over K of degree (say) m; then $m | n$ and our extension is even cyclic with generating automorphism σ such that $^\sigma(\sqrt[r]{a}) = \zeta^{rp} \sqrt[r]{a}$ ($r := \frac{n}{m}$).

Moreover, $\sigma(\sqrt[m]{a}) = \sigma((\sqrt[m]{a})^m) = \zeta^{rm}(\sqrt[m]{a})^m = \sqrt[m]{a}$, hence $K(\sqrt[m]{a}) = K$, and therefore we obtain from Lemma 9 and Lemma 2 - note $\sqrt[m]{\sqrt[n]{a}} = \sqrt[m]{a}$ - the important

Theorem 2. *Assume* $\langle \zeta \rangle = \mu_n \subset K$ *and write* $m := |K(\sqrt[n]{a}):K|$ *(* $a \in K^*$ *);
then* $r := \frac{n}{m} \in \mathbb{N}$ *and we have the equation*
$$[a,b;n,K,\zeta] = [b^{-1}, K(\sqrt[m]{a})/K, \sigma] \quad in \quad Br(K)$$
where $\sigma(\sqrt[m]{a}) = \zeta^r \sqrt[m]{a}$. □

Theorem 1 in §10., Theorem 2 and Corollary 3 clearly imply

Corollary 4. *One has* $[a,b;n,K] = 0$ *in* $Br(K)$ *if and only if* b *is a norm for the extension* $K(\sqrt[m]{a})/K$ *which is the case if and only if* a *is a norm for the extension* $K(\sqrt[n]{b})/K$. □

An important application of Corollary 4 is

Lemma 10. *If* $a,b \in K^*$ *such that* $a + b = c^n$ *for some* $c \in K$, *then*
$$[a,b;n] = 0 \quad in \quad Br(K).$$

Proof. Write $L := K(\sqrt[n]{a})$, $\Gamma := Gal(L/K) = \sigma$ where $\sigma(\sqrt[m]{a}) = \zeta^r \sqrt[m]{a}$
($r = \frac{n}{m}$, $m = |L:K|$; cf. the remarks previous to Theorem 2 above). Then
$$T^n - a = \prod_{i=0}^{n-1}(T - \zeta^i \sqrt[n]{a}) \in L[T], \text{ hence (set } T = c \text{)}$$
$$b = c^n - a = \prod_{i=0}^{n-1}(c - \zeta^i \sqrt[n]{a}) = \prod_{j=0}^{m-1}(\prod_{\rho=0}^{r-1}(c - \zeta^{rj+\rho}\sqrt[n]{a})) =$$
$$= \prod_{j=0}^{m-1} \sigma^j (\prod_{\rho=0}^{r-1}(c - \zeta^\rho \sqrt[n]{a})) = N_{L/K}(\prod_{\rho}(c - \zeta^\rho \sqrt[n]{a})). \quad □$$

Now consider the special cases $c = 1$ and $c = 0$ in Lemma 10:

Corollary 5. $[a,1-a] = 0$ *and* $[a,-a] = 0$. □

Lemma 4 and Corollary 5 imply then

Lemma 11. $[a,b] = [-\frac{a}{b}, a+b]$, $[a,a-b] = [b, \frac{a-b}{a}] + [a,-1]$ *and*
$$[a,a;n] = [a,-1;n] \begin{matrix} \in {}_2Br(K) & \text{if } n \text{ is even}; \\ = 0 & \text{if } n \text{ is odd}. \end{matrix}$$

Proof. $[a,a] = [a,(-a)(-1)] = [a,-a] + [a,-1]$ and, if n is odd: $[a,-1;n] = [a,(-1)^n;n] = n[a,-1;n] = 0$. Moreover, if $c := a + b$, then $0 = [ac^{-1}, 1-ac^{-1}] = [ac^{-1}, bc^{-1}] = [a,b] - [a,c] - [c,b] + [c,c] =$
$= [a,b] - ([a,c] - [b,c] + [-1,c]) = [a,b] - [-\frac{a}{b}, a+b]$ and consequently
$[a,a-b] = [a,-1] + [a,b-a] = [a,-1] + [\frac{-a}{b-a}, b] = [a,-1] + [b, \frac{a-b}{a}]$. □

Definition 3. *Let* K *be a commutative field, then we define*
$$K_2(K) := K^* \otimes_{\mathbb{Z}} K^* / \langle a \otimes (1-a) \mid a \in K^* \rangle.$$
We write $K_2(K)$ *additively and denote by* $\{a,b\}$ *the class of* $a \otimes b$ *in*

the group $K_2(K)$. An element $\{a,b\}$ is called a "symbol", i.e. $K_2(K)$ is generated by symbols.

Thanks to Corollaries 2/5 we have

Theorem 3. Assume $\mu_n \subset K$, then there is exactly one homomorphism
$$R_{n,K}: K_2(K)/nK_2(K) \longrightarrow {}_n Br(K) \otimes_Z \mu_n$$
such that $\{a,b\} \mapsto [a,b;n,K,\zeta] \otimes \zeta$. This homomorphism is independent of the choice of the primitive n-th root of unity ζ. □

Definition 4. The homomorphism $R_{n,K}$ from Theorem 3 is called the "(abstract) norm residue homomorphism".

Later in §17. we shall see that $R_{n,K}$ is always an isomorphism ! Moreover, we should point out that in general $K_2(R)$ may be defined *for any* ring R (with $1 \neq 0$); for this see J. Milnor [1971]. In the case of a commutative field Matsumoto's Theorem (cf. §§11/12. *op. cit.*) leads to our definition. It is worth mentioning that Matsumoto's Theorem can be carried over *mutatis mutandis* to the general skew field case (cf. U. Rehmann [1978]). See §17. for more information about the functor K_2.

Hitherto we have worked under the assumption $\mu_n \subset K$ (which implies $\mathrm{char}(K) \nmid n$). Now let us interrupt our investigations for a few remarks concerning the case of cyclic algebras of reduced degree $p := \mathrm{char}(K) \neq 0$.

So in what follows assume $p := \mathrm{char}(K) \neq 0$ and consider the additive homomorphism
$$\wp: K^+ \to K^+ , \quad \wp x := x^p - x$$
(this notation, which is nowadays the standard notation, goes back to E. Witt) which gives rise to the exact sequence

(12) $\qquad 0 \longrightarrow F_p \overset{\subseteq}{=\!=\!=} K^+ \xrightarrow{\wp} K^+$.

Now choose $a \in K^*$, $b \in K$ and select an auxiliary field extension L/K such that $b \in \wp K$ (for instance take L a separable closure of K). Moreover consider the matrices

$A \in M_p(K)$ according to (3) in §10., and $B := \begin{pmatrix} \ddots & & 0 \\ & \bar{b}+i & \\ 0 & & \ddots \end{pmatrix} \in$

$\in M_p(L)$ where $\bar{b} \in L$ is
such that $\wp \bar{b} = b$. $\quad\uparrow$
i-th row and column

Lemma 1 in §2. yields

(13) $\qquad A^p = a$, $\wp B = b$, and $AB = (B+1)A$

as well as the following consequences thereof:

(14) $\quad A^i B A^{-i} = B + i \quad$ and $\quad A B^j A^{-1} = (B+1)^j \quad (\ i,j \in \mathbb{Z}\ )$.

Now consider the n^2 matrices $A^i B^j$; in precisely the same way as in the case discussed previous to Theorem 1 we can show that they are L-linearly independent (the only difference that matters is that the Vandermonde matrix in question has now entries $v_{ij} = (\flat+i)^{j-1}$ ($1 \le i,j \le p$)).
Thus we get (cf. proof of Theorem 1)

(15) $\quad \bigoplus_{i=0}^{n-1} \bigoplus_{j=0}^{n-1} K A^i B^j =: A \qquad$ (note $A \subseteq M_p(L)$)

and

Theorem 4. *If* $p := \mathrm{char}(K) \ne 0$, *then all K-algebras* A *generated by two elements* a, b *such that*

(16) $\quad a := a^p \in K^*$, $b := \wp b \in K$, $ab = (b+1)a$

are isomorphic (namely $\simeq A$ *in (15)). They are central simple K-algebras (more precisely: either a K-skew field or* $\simeq M_p(K)$ *) and any field* L *such that* $K \subseteq L$ *and* $b \in \wp L$ *is a splitting field.* □

Definition 5. *We denote any of the isomorphic K-algebras described in Theorem 1 by*

$(a, b; p, K)$

and call it again a "power norm residue algebra". Moreover, we write

$[a, b; p, K) := [(a, b; p, K)] \in \mathrm{Br}(K)$.

If no confusion can arise we omit any of the symbols p, K *in* $(a, b; p, K)$ *and* $[a, b; p, K)$.

Definition 6. *If* $\mathrm{char}(K) = 2$ *we write*

$\left[\dfrac{a,b}{K}\right]$ *in place of* $(a,b;2,K)$

and call such an algebra a "quaternion algebra".

See §14. for more details on quaternion algebras.

Lemma 12. *Let* $p := \mathrm{char}(K) \ne 0$, $|K(\sqrt[p]{a}):K| = p$ *and* $[A] \in \mathrm{Br}(K(\sqrt[p]{a})/K)$, *then*

$[A] = [a, b; p, K)$ *in* $\mathrm{Br}(K)$ *for suitable* $b \in K$.

Proof. If $[A] = 0$ take any $b \in \wp K$ (e.g. $b = 0$). If $[A] \ne 0$, then replace A by D such that $[A] = [D]$, $a := \sqrt[p]{a} \in D$ and $|D:K| = p^2$ (cf. Theorem 7 in §9.). Then D is a skew field (cf. Theorem 1(*f*) in §9.) and Lemma 6 in §9. (see also its proof) implies the existence of an element $b \in D$ such that $aba^{-1} = b + 1$. Now, if $L := K(b)$, for reasons of degree (cf. Theorem 4 in §7.) L/K is cyclic of degree p with generating

automorphism σ: b ↦ b+1 . It follows (cf. (12))
$$^\sigma(\wp b) = (b+1)^p - (b+1) = b^p - b = \wp b ,$$
hence b := $\wp b \in \text{Fix}_L(\sigma)$ = K . Therefore - again for reasons of degree - Theorem 4 implies
$$[a,b;p,K> = [D] = [A] . \quad \square$$

Lemma 13 . $(a,b> \otimes_K (a',b'> \simeq (aa',b> \otimes_K (a',b'-b> , i.e. [a,b> + [a',b'> =$
$= [aa',b> + [a',b'-b>$ in Br(K).

Proof. *mutatis mutandis* identical with the proof of Lemma 3 (cf. Exercise 1). □

Now, similar to the reasoning which led from Lemma 3 via (7),..,(11) to Lemma 4, we may conclude

Lemma 14 . *Let* p := char(K) ≠ 0 , *then the assignment* (a,b) ↦
↦ [a,b;p,K> *induces a Z-bilinear map (multiplicative in the first and additive in the second argument)*
$$K^*/(K^*)^p \times K/\wp K \to {}_p Br(K) . \quad \square$$

Now, using the notation introduced in Exercise 1 in §9., it is clear from §10. and Theorem 4 (cf. also Exercise 1):

Theorem 5 . *Assume* p := char(K) ≠ 0 , *then*
$$[a,b;p,K> = [a,K(\tfrac{1}{\wp}b)/K,\sigma] \quad in \quad Br(K)$$
with the convention $K(\tfrac{1}{\wp}b)$ = K *and* σ = id *if* b ∈ $\wp$K . □

Theorem 1 in §10. implies then

Corollary 6 . *One has* [a,b;p,K> = 0 *in* Br(K) *if and only if* a *is a norm for the extension* $K(\tfrac{1}{\wp}b)/K$. □

Lemma 15 . *If* a ∈ K* *and* c ∈ K , *then* [a,c^{p}a;p> = 0 *in* Br(K) .

Proof. see Exercise 1 . □

Now we turn our attention back to the case where $\mu_n \subset K$ (which implies char(K)∤n). Here we study the problem to what extent power norm residue algebras remain at least (up to ~) tensor products of cyclic algebras under *corestriction* (compare with Lemma 8). This turns out to be a difficult problem and only partial results are available nowadays; we shall discuss some of these and begin with some preparatory remarks.

Lemma 16 . *Let* L/K *be finite Galois* , Γ := Gal(L/K) *and* $\mu_n \subset K$; *then* (σ ∈ Γ , μ_n = ⟨ζ⟩)

$$^\sigma(a,b;n,L,\zeta) \simeq (^\sigma a,^\sigma b;n,L,^\sigma \zeta) \quad , \text{ i.e.}$$

$$^\sigma[a,b;n,L,\zeta] = [^\sigma a,^\sigma b;n,L,^\sigma \zeta] \quad \text{in } Br(K).$$

Proof. Thanks to Lemma 2 in §8. and Lemma 1 in §9. it suffices to establish a ring homomorphism

$$f: A' := (^\sigma a,^\sigma b;n,L,^\sigma \zeta) \to (a,b;n,L,\zeta) =: A$$

such that $f(x) = \sigma^{-1}(x)$ for all $x \in L$. Now assume (cf. Theorem 1) $A = \langle a,b \mid a^n = a, b^n = b, ab = \zeta ba \rangle$ and $A' = \langle a',b' \mid a'^n = ^\sigma a, b'^n = ^\sigma b, a'b' = ^\sigma \zeta b'a' \rangle$, then it is obvious that the assignment

$$a' \mapsto a, \quad b' \mapsto b, \quad x \mapsto \sigma^{-1}(x) \quad (x \in L)$$

induces the required homomorphism $f: A' \to A$. □

Theorem 6. *Let p be a prime $\neq char(K)$, $\mu_p = \langle \zeta \rangle$, $L := K(\zeta)$, $a \in L^*$, $b \in K^*$ and $D := (a,b;p,L,\zeta)$ a skew field; then either $c_{L/K}[D] = 0$ or there exists a cyclic extension F/K of degree p with generating automorphism γ such that*

$$c_{L/K}[D] = [b,F/K,\gamma] \quad \text{in } Br(K).$$

Proof. Clearly L/K is cyclic of degree m with generating automorphism δ ; here $m|p-1$, $^\delta\zeta = \zeta^s$ for some $s \neq 0 \mod p$ such that $s^m \equiv 1 \mod p$. Now set $G := Gal(L/K) = \langle \delta \rangle$ and consider (cf. Lemma 4 in §9.)

$$N_G[D] = \sum_{i=0}^{m-1} {}^{\delta^i}[D] = \sum_{i=0}^{m-1} {}^{\delta^i}[a,b;\zeta] = \sum_{i=0}^{m-1} [^{\delta^i}a,b;^{\delta^i}\zeta] =$$

$$= \sum_{i=0}^{m-1} [^{\delta^i}a,b;\zeta^{s^i}] = \sum_{i=0}^{m-1} [(^{\delta^i}a)^{s^{m-i}},b;\zeta] = [c,b;p,L,\zeta]$$

where

$$c := \prod_{i=0}^{m-1} (^{\delta^i}a)^{s^{m-i}} \in L^*.$$

Here we have made use of Lemma 16 as well as Lemma 5. A straightforward calculation shows (cf. Lemma 4)

$$^\delta c \equiv (^\delta c)^{s^m} = \prod_{i=0}^{m-1} (^{\delta^{i+1}}a)^{s^{m-(i+1)} \cdot s^{m+1}} = (ca^{s^m-1})^{s^{m+1}} \equiv$$

$$\equiv c^s \mod (L^*)^p, \quad \text{i.e.}$$

(17) $\quad ^\delta c = c^s d^p$ for some $d \in L^*$.

Now consider the field $N := L(\sqrt[p]{c}) = K(\zeta,\sqrt[p]{c})$. Obviously we have *either* $N = L$ (i.e. $c \in (L^*)^p$ and therefore $N_G[D] = 0$ thanks to Lemma 4) *or* N/L is cyclic of order p with generating automorphism τ defined by $\tau(\sqrt[p]{c}) = \zeta\sqrt[p]{c}$ (cf. Field Theory). By virtue of (17) it is feasible to extend the K-automorphism δ of L to a K-automorphism σ of N via

$$\sigma(\sqrt[p]{c}) := (\sqrt[p]{c})^s d \quad (\text{ d as in (17)}) \ , \ ^\sigma \zeta = \zeta^s \ .$$

Consequently N/K is Galois with $\Gamma := \text{Gal}(N/K) = \langle \sigma, \tau \rangle$; here $^{\sigma\tau}\zeta =$
$= \zeta^s = {}^{\tau\sigma}\zeta$ and $^{\sigma\tau}(\sqrt[p]{c}) = {}^\sigma(\zeta\sqrt[p]{c}) = \zeta^s(\sqrt[p]{c})^s d = {}^\tau((\sqrt[p]{c})^s d) = {}^{\tau\sigma}(\sqrt[p]{c})$, hence Γ is abelian and even cyclic with generator $\sigma\tau$ (thanks to $|\Gamma| = mp$ with coprime m and p). If we denote by F *the* intermediate field of N/K of degree p over K - i.e. $F = \text{Fix}_N(\sigma)$ -, then there exists a generating automorphism γ of the cyclic extension F/K such that $\gamma^{-1} = \sigma\tau|_F$. It follows (cf. Lemma 2 and Lemmas 3/6/8 in §10.)

$$[c,b;p,L,\zeta] = [b,N/L,\tau^{-1}] = [b^m,N/L,\tau^{-m}] = [b^m,N/L,(\sigma\tau)^{-m}] =$$
$$= r_{L/K}[b^m,N/K,(\sigma\tau)^{-1}] = r_{L/K}[b,F/K,(\sigma\tau)^{-1}|_F] = r_{L/K}[b,F/K,\gamma] \ .$$

On the other hand we know $r_{L/K}(c_{L/K}[D]) = N_G[D] = [c,b;p,L,\zeta]$ (cf. Theorem 5 in §9.), hence, since $r_{L/K}$ is *injective* on $_p\text{Br}(K)$ thanks to Corollary 13 in §9., we have *either* $c_{L/K}[D] = 0$ *or* $c_{L/K}[D] = [b,F/K,\gamma]$ in $\text{Br}(K)$. □

An easy consequence of the theorem above is

Albert's Criterion . *Let p be a prime and D a K-skew field of index p ; then D is isomorphic to a cyclic algebra $(b,F/K,\gamma)$ if and only if D contains a field $K(\sqrt[p]{b})$ such that $|K(\sqrt[p]{b}):K| = p$.*

Proof. The "only if" is obvious and always (i.e. for arbitrary index) true. As for the "if" part, the case $p = \text{char}(K)$ has been settled in Lemma 12 (note Theorem 5 and Corollary 5 in §9.). Now assume $p \neq \text{char}(K)$, let $\langle\zeta\rangle = \mu_p$ and set $L := K(\zeta)$; then $A := D \otimes_K L$ is an L-skew field of index p (see Corollary 8 in §9.) such that
$$K(\sqrt[p]{b}) \otimes_K L \simeq L(\sqrt[p]{b}) \subseteq A \quad \text{and} \quad |L(\sqrt[p]{b}):L| = p$$
(note: L/K is Galois and $p, |L:K|$ are coprime). Lemma 1 and Lemma 3(β)/ Corollary 5 in §9. imply
$$A \simeq (a,b;p,L,\zeta) \quad \text{for some} \quad a \in L^* \ ,$$
hence (because of Theorem 5 in §9. and Theorem 6 above), if $m := |L:K|$,
$$m[D] = c_{L/K}(r_{L/K}[D]) = c_{L/K}[A] = [b,F/K,\gamma]$$
for some cyclic extension F/K of degree p (note that $m[D] = c_{L/K}[A] = 0$ would imply $[D] = 0$ (which contradicts our assumptions) because m and p are coprime (cf. Corollary 7 in §9.)). Therefore
$$D \simeq (b,F/K,\gamma^m)$$
because of Lemma 3(β) in §9. and Lemmas 3/4 in §10. □

The following result is of utmost importance.

Theorem 7. *Let* L/K *be finite separable,* $\langle \zeta \rangle = \mu_n \subset K$, $a \in L^*$ *and* $b \in K^*$, *then*
$$c_{L/K}[a,b;n,L,\zeta] = [N_{L/K}(a),b;n,K,\zeta] \ .$$

Proof. Let N/K be finite Galois such that L is an intermediate field of N/K, $\Delta := \text{Gal}(N/L) \leq \text{Gal}(N/K) =: \Gamma$ and $\Gamma = \bigcup_{\rho \in R} \rho\Delta$. Then, because of Lemma 3 (use it repeatedly with $b' = b$) and Lemmas 8/16 we find (in the notation of §8.)

$$(a,b;L,\zeta)^{(\Gamma:\Delta)} = \bigotimes_{\rho \in R} {}^\rho((a,b;L,\zeta) \otimes_L N) \simeq \bigotimes_{\rho \in R} ({}^\rho a,b;N,\zeta) \simeq$$
(18)
$$\simeq (\prod_{\rho \in R} {}^\rho a,b;N,\zeta) \otimes_N (\bigotimes_{\Delta \not\ni \rho \in R} ({}^\rho a,1;N,\zeta)) \simeq$$
$$\simeq \{(N_{L/K}(a),b;K,\zeta) \otimes_K M_m(K)\} \otimes_K N \quad (\ m := n^{|L:K|-1}\ )$$

An analysis of the proof of Lemma 3 shows that Γ acts trivially on the left factor of the right hand side of (18) (cf. Exercise 2 ; the action on the left hand side of (18) is via (11) in §8.), hence our claim is obvious thanks to Theorems 1/2 in §6. □

Lemma 17. *Let* $p := \text{char}(K) \neq 0$, $\langle \zeta \rangle = \mu_n \subset K$ *and assume* L/K *finite separable. Consider monic irreducible polynomials* $f,g \in K[T]$ *such that* g *is separable with* $f(T) = g(T^{p^e})$ *for some* e. *Now choose* t *such that* $tp^e \equiv 1 \mod n$, *then, for any* $a \in L^*$ *we have the equation*
$$[a,f(a);n,L,\zeta] = t[a^{p^e},g(a^{p^e});n,L,\zeta] \quad in \quad \text{Br}(L) \ .$$

Proof. Clear because of Lemma 4. □

The next result is due to J. Tate [so far unpublished] and the author is very grateful for J. Tate's permission to publish his argument here. His ideas, incidentally, are refinements of arguments due to S. Rosset [1977], [a].

Tate's Reciprocity Lemma. *Assume* $\langle \zeta \rangle = \mu_n \subset K$ *and let* L/K *resp.* F/K *be finite separable extensions of degrees* $d := |L:K|$ *resp.* $r := |F:K|$. *Write* $L = K(a)$ *resp.* $F = K(c)$ *and choose monic irreducible polynomials* $p,f \in K[T]$ *such that* $p(a) = 0 = f(c)$. *Then we have the "reciprocity law"*
$$c_{L/K}[a,f(a);n,L,\zeta] = c_{F/K}[c,\frac{p(c)}{p(0)};n,F,\zeta] + [N_{L/K}(a),(-1)^r;n,K,\zeta] \ .$$

Proof. Let N be a common splitting field of f and p over K, hence $a,c \in N$. Now let $f = \prod_{i=1}^{s} f_i$ be the factorisation of f into monic irreducible factors $f_i \in L[T]$ of degree r_i ($\sum_{i=1}^{s} r_i = r$). Moreover, for each i let c_i be a root of f_i and put $N_i := K(a,c_i) = L(c_i)$. Now choose $\sigma_i \in \text{Gal}(N/K)$ such that $\sigma_i(c_i) = c$ and define $a_i := \sigma_i(a)$ in N as well as $M_i := \sigma_i(N_i) = K(a_i,c) = F(a_i)$. Then, if we denote by $p_i \in F[T]$ the monic irreducible polynomial such that $p_i(c_i) = 0$ (i = = 1,..,s), we know from Field Theory that we get $p = \prod_{i=1}^{s} p_i$. Now it is quite clear from §8. (cf. Exercise 2) that we have the equalities

(19)
$$c_{M_i/K}[c,\frac{c-a_i}{-a_i};M_i] = c_{\sigma_i(N_i)/K}[\sigma_i(c_i),\frac{\sigma_i(c_i)-\sigma_i(a)}{-\sigma_i(a)};\sigma_i(N_i)] =$$
$$= c_{N_i/K}[c_i,\frac{c_i-a}{-a};N_i] \quad \text{in Br(K)} \quad (i = 1,...,s) ,$$

hence (cf. Lemma 11)

(20)
$$c_{N_i/K}[a,a-c_i;N_i] = c_{N_i/K}[a,-1;N_i] = c_{M_i/K}[c,\frac{c-a_i}{-a_i};M_i] \quad \text{in Br(K)}.$$

Using Lemma 4, Theorem 7 and Lemma 11 in §8. we compute

(21)
$$c_{L/K}[a,f(a);L] = c_{L/K}(\sum_{i=1}^{s} [a,f_i(a);L]) =$$
$$= c_{L/K}(\sum_{i=1}^{s} [a,N_{N_i/L}(a-c_i);L]) = \sum_{i=1}^{s} c_{N_i/K}[a,a-c_i;N_i] ,$$

(22)
$$[N_{L/K}(a),(-1)^r;K] = c_{L/K}[a,(-1)^r;L] = \sum_{i=1}^{s} c_{L/K}[a,(-1)^{r_i};L] =$$
$$= \sum_{i=1}^{s} c_{N_i/K}[a,-1;N_i]$$

and

(23)
$$c_{F/K}[c,\frac{p(c)}{p(0)};F] = c_{F/K}(\sum_{i=1}^{s} [c,\frac{p_i(c)}{p_i(0)};F]) =$$
$$= c_{F/K}(\sum_{i=1}^{s} [c,N_{M_i/F}\left(\frac{c-a_i}{0-a_i}\right);F]) = \sum_{i=1}^{s} c_{M_i/K}[c,\frac{c-a_i}{-a_i};M_i] .$$

Now our reciprocity law follows from (20),..,(23) . □

An immediate application of Tate's Reciprocity Lemma is the

Rosset-Tate Theorem. *Assume* $\langle\zeta\rangle = \mu_n \subset K$ *and let* L/K *be a finite separable extension of degree* d , *then*

$$c_{L/K}[a,b;n,L,\zeta] = \sum_{j=1}^{d} [a_j,b_j;n,K,\zeta] \text{ for suitable } a_j,b_j \in K^*.$$

Proof. We proceed by induction on the degree $d = |L:K|$; the case d = 1 is trivial. Now assume d > 1 . Thanks to Lemma 11 in §8. our claim is transitive, hence we may assume that the extension L/K has no proper intermediate field. Moreover, by virtue of Corollary 3 and Theorem 7 it suffices to assume a $\notin$ K* which amounts to L = K(a) (see above) and therefore

(24) $\quad$ b = b'f(a) with a monic irreducible polynomial f ∈ K[T]
$\quad\quad\quad\quad$ of degree r < d and some b' ∈ K* .

Let p ∈ K[T] be the (monic) minimal polynomial of a over K , c a root of f and F := K(c) . From Field Theory and Lemma 17 we know that, after possible replacement of a by a^{p^e}, we may assume f to be even separable, hence

(25) $\quad$ F/K is separable and $|F:K| < |L:K|$.

Now Tate's Reciprocity Lemma, Theorem 7 and Lemma 4 imply

$$c_{L/K}[a,b;L] = c_{L/K}[a,b';L] + c_{L/K}[a,f(a);L] =$$

$$= c_{L/K}[a,b';L] + [N_{L/K}(a),(-1)^r;K] + c_{F/K}[c,\frac{p(c)}{p(0)};F] =$$

$$= [N_{L/K}(a),b'(-1)^r;K] + c_{F/K}[c,\frac{p(c)}{p(0)};F] ,$$

hence our claim by induction hypothesis (see (25)). □

A nice application of the Rosset-Tate Theorem is

Rosset's Theorem . *Let* p *be a prime* $\neq$ char(K) , *assume* $\langle\zeta\rangle = \mu_p \subset K$, *and let* D *be a K-skew field of index* p ; *then*

$$[D] = \sum_{j=1}^{d} [a_j,b_j;p,K,\zeta] \text{ for suitable } a_j,b_j \in K^*, d \leq (p-1)! ,$$

hence D *has the abelian splitting field* $K(\sqrt[p]{a_1},..,\sqrt[p]{a_d})$ *over* K .

Proof. Let M be a maximal commutative subfield of a given K-skew field D of index p , hence $|M:K| = p$ and $[D] \in Br(M/K)$ (cf. Theorem 4 in §7. and Corollary 5 in §9.). Now let N be the Galois closure of M over K , $\Gamma := Gal(N/K)$, Γ_p a p-Sylow subgroup of Γ and $L := Fix_N(\Gamma_p)$ the corresponding p-Sylow subfield of N/K . Set $d := |L:K|$, then

$\quad\quad$ N/L is cyclic of degree p and $d|(p-1)!$, hence d and p
$\quad\quad$ are coprime, in particular $td \equiv 1 \mod p$ for some t .

Thanks to $[D] \in Br(M/K) \subseteq Br(N/K)$ we find $r_{L/K}[D] \in Br(N/L)$, hence Corollary 1 implies

$$r_{L/K}[D] = [a,b;p,L,\zeta] \text{ in } Br(L) \text{ for some } a,b \in L^* .$$

Now apply $c_{L/K}$ to both sides of the above equation; using Theorem 5 in §9., Lemma 4 and the Rosset-Tate Theorem we obtain at ease

$$[D] = td[D] = c_{L/K}(r_{L/K}(t[D])) = c_{L/K}[a,b^t;p,L,\zeta] =$$
$$= \sum_{j=1}^{d} [a_j,b_j;p,K,\zeta] \text{ for suitable } a_j,b_j \in K^* ,$$

hence $K(\sqrt[p]{a_1},..,\sqrt[p]{a_d})$ is a splitting field of D because of Theorem 2. □

Now let p be a prime $\neq \text{char}(K)$ and $\mu_p = \langle \zeta \rangle$. Then $K(\zeta)/K$ is cyclic of degree dividing $p-1$, so any elementary p-abelian extension of $K(\zeta)$ can be enlarged to an elementary p-abelian extension $L/K(\zeta)$ such that L/K is also Galois and then *a fortiori* soluble. So we have proved the important

Corollary 7. *Let p be a prime $\neq \text{char}(K)$, then any K-skew field of index p has a metabelian (and hence soluble) splitting field over* K. □

As we shall see later in §15. the assumption $p \neq \text{char}(K)$ in the above corollary is superfluous.

Exercise 1. Give detailed proofs of Lemmas 13/14/15 and Theorem 5.

Exercise 2. Give detailed proofs of Theorem 7 and the second equation in (19).

Exercise 3. Prove

Dickson's Theorem. *Every K-skew field of index 3 is a cyclic algebra.*
(cf. A.A. Albert [1939,p.177]; use Albert's Criterion)

Added in proof. Concerning Tate's Reciprocity Lemma cf. also the recent paper

S. Rosset, J.Tate

[1982]   A Reciprocity Law for K_2-Traces, Forschungsinstitut für Mathematik ETH Zürich, *Preprint* (September 1982)

§ 12 . BRAUER GROUPS AND GALOIS COHOMOLOGY

In this paragraph we shall achieve interesting results by combining Theorem 5 in §7. with Theorems 7/9 in §9.

Theorem 1. *Let* L/K *be finite Galois,* $\Gamma := \text{Gal}(L/K)$, $[A] \in \text{Br}(L/K)$, $L \subseteq A$ *and* $|L:K|^2 = |A:K|$. *Then there exist elements* $e_\sigma \in A^*$ ($\sigma \in \Gamma$) *such that*

(i) $\qquad A = \bigoplus_{\sigma \in \Gamma} L e_\sigma$, $e_{\text{id}} = 1$, $e_\sigma x = {}^\sigma\!x e_\sigma$ ($x \in L$, $\sigma \in \Gamma$)

and

(ii) $\qquad e_\sigma e_\tau = x(\sigma,\tau) e_{\sigma\tau}$ ($\sigma,\tau \in \Gamma$, $x(\sigma,\tau) \in L^*$).

Here

(iii) $\qquad x(\sigma,\tau) x(\sigma\tau,\rho) = {}^\sigma\!x(\tau,\rho) x(\sigma,\tau\rho)$ *and* $x(\sigma,\text{id}) = 1 = x(\text{id},\tau)$.

Moreover, if there are elements f_σ *satisfying (i) just as the elements* e_σ *do, and if those* f_σ *define elements* $y(\sigma,\tau)$ *according to (ii), then*

(iv) $\qquad e_\sigma = z(\sigma) f_\sigma$ ($\sigma \in \Gamma$) *where* $z(\sigma) \in L^*$, $z(\text{id}) = 1$

and

(v) $\qquad \dfrac{x(\sigma,\tau)}{y(\sigma,\tau)} = \dfrac{z(\sigma)^\sigma z(\tau)}{z(\sigma\tau)}$ ($\sigma,\tau \in \Gamma$).

Proof. *(i)* is clear from Theorem 5 in §7. thanks to $Z_A(L) = L$ (cf. Theorem 3 in §7.). Now we have

$$e_{\sigma\tau} x e_{\sigma\tau}^{-1} = {}^{\sigma\tau}\!x = {}^\sigma({}^\tau\!x) = e_\sigma e_\tau x e_\tau^{-1} e_\sigma^{-1} \text{ for all } x \in L \text{ , hence}$$

$$e_{\sigma\tau}^{-1} e_\sigma e_\tau \in Z_A(L) = L \text{ (see above)}$$

and therefore *(ii)* with $x(\sigma,\tau) := {}^{\sigma\tau}(e_{\sigma\tau}^{-1} e_\sigma e_\tau)$. *(iii)* merely reflects the law of associativity in A, namely

$$x(\sigma,\tau) x(\sigma\tau,\rho) e_{\sigma\tau\rho} = x(\sigma,\tau) e_{\sigma\tau} e_\rho = (e_\sigma e_\tau) e_\rho = e_\sigma (e_\tau e_\rho) =$$
$$= e_\sigma x(\tau,\rho) e_{\tau\rho} = {}^\sigma\!x(\tau,\rho) e_\sigma e_{\tau\rho} = {}^\sigma\!x(\tau,\rho) x(\sigma,\tau\rho) e_{\sigma\tau\rho} ,$$

hence *(iii)* since the elements $e_{\sigma\tau\rho}$ are invertible (the rest of *(iii)* is obvious because of *(ii)* and $e_{\text{id}} = 1$). Now *(iv)*: again

$$e_\sigma x e_\sigma^{-1} = {}^\sigma\!x = f_\sigma x f_\sigma^{-1} \text{ for all } x \in L \text{ , hence}$$

$f_\sigma^{-1} e_\sigma \in Z_A(L) = L$ (see above)

and therefore *(iv)* with $z(\sigma) := (f_\sigma^{-1} e_\sigma) \in L$. Finally, *(v)* results from a simple calculation using *(i),(ii)* and *(iv)*. □

The formulae *(ii),..,(v)* are *one* historical source of a series of definitions, namely (cf. §6.): let M be a left Γ-module, then

$$C^2(\Gamma,M) := \{ x: \Gamma \times \Gamma \to M \mid \begin{array}{l} x(\sigma,\tau) + x(\sigma\tau,\rho) = {}^\sigma x(\tau,\rho) + x(\sigma,\tau\rho) \ \& \ x(\sigma,1) = 0 = x(1,\tau) \end{array} \}$$

is called the set of *2-cocycles (of* Γ *with values in* M *)*. C^2 carries the structure of a Z-module (by pointwise definition of the (say) addition) and an easy calculation shows that the 2-*coboundaries*

$$B^2(\Gamma,M) := \{ x: \Gamma \times \Gamma \to M \mid \begin{array}{l} x(\sigma,\tau) = z(\sigma) - z(\sigma\tau) + {}^\sigma z(\tau) \\ \text{where } z: \Gamma \to M \text{ with } z(1) = 0 \end{array} \}$$

form a Z-submodule of C^2. Note that in more old-fashioned terminology C^2 resp. B^2 is called the group of *factor sets* resp. *principal factor sets (from* Γ *into* M *).*

Definition 1. *Let* Γ *be a group and* M *a left* Γ-*module, then* $H^2(\Gamma,M) := C^2(\Gamma,M)/B^2(\Gamma,M)$ *is called the " 2nd Cohomology Group of* M *"*.

Theorem 2. *Let* L/K *be finite Galois,* $\Gamma := \mathrm{Gal}(L/K)$, $n := |L:K| = |\Gamma|$ *and* $x \in C^2(\Gamma,L^*)$. *Define on the* n^2-*dimensional* K-*vector space*

$$(x,L/K) := \bigoplus_{\sigma \in \Gamma} L e_\sigma$$

a multiplication by the following formulae

$$e_\sigma x = {}^\sigma x e_\sigma \ (x \in L), \ e_\sigma e_\tau = x(\sigma,\tau) e_{\sigma\tau} \text{ and}$$
$$(\sum_{\sigma \in \Gamma} x_\sigma e_\sigma)(\sum_{\tau \in \Gamma} y_\tau e_\tau) = \sum_{\sigma,\tau \in \Gamma} x_\sigma e_\sigma y_\tau e_\tau \ .$$

Then the elements e_σ *are invertible,* $e_{id} = 1$, *and* (x,L/K) *is a central simple* K-*algebra with splitting field* L, *i.e.*

$$[x,L/K] := [(x,L/K)] \in \mathrm{Br}(L/K) \ .$$

Proof. Write $A := (x,L/K)$; the associativity of the multiplication defined above is clear (same calculation as in the proof of formula *(iii)* in Theorem 1); so is $e_{id} = 1$ (thanks to $x(\sigma,id) = 1 = x(id,\tau)$) and hence

(1) $\quad e_\sigma^{-1} = x(\sigma^{-1},\sigma)^{-1} e_{\sigma^{-1}} = ({}^{\sigma^{-1}}x(\sigma,\sigma^{-1}))^{-1} e_{\sigma^{-1}}$.

Now embed $L \hookrightarrow A$ via $x \mapsto x e_{id} = x$, and assume

$$0 = x(\sum_{\sigma \in \Gamma} x_\sigma e_\sigma) - (\sum_{\sigma \in \Gamma} x_\sigma e_\sigma) x = \sum_{\sigma \in \Gamma} (x - {}^\sigma x) x_\sigma e_\sigma \ (x \in L) ,$$

This implies first $Z_A(L) = L$ (choose $x \in L$ such that ${}^\sigma x \neq x$ for all

$\sigma \neq \mathrm{id}$), hence $Z(A) = Z_A(A) \subseteq Z_A(L) = L$ (cf. (5) and (8) in §7.) and therefore $Z(A) = \mathrm{Fix}_L(\Gamma) = K$. Now we must show that A has no proper two-sided ideal (cf. Theorem 1(b) in §9.); this is done *mutatis mutandis* as in case of the corresponding statement in the course of the proof of Lemma 5 in §7. (here we have $B = L$ and e_σ in place of e^i; moreover, L/K is not cyclic any more, however, this does not affect our arguments here). Finally we conclude $[x, L/K] \in \mathrm{Br}(L/K)$ thanks to Theorem 7 in §9. □

Definition 2. *An algebra* $(x, L/K)$ *as in Theorem 2 is called a "crossed product".*

Lemma 1. *In the situation of* Theorem 2 *let* $x, y \in C^2(\Gamma, L^*)$ *be such that* $x \equiv y \mod B^2(\Gamma, L^*)$, *then*

$$(x, L/K) \simeq (y, L/K) .$$

Proof. By assumption we have

$$\frac{x(\sigma, \tau)}{y(\sigma, \tau)} = \frac{z(\sigma)^\sigma z(\tau)}{z(\sigma\tau)}$$ for some function $z: \Gamma \to L^*$ where $z(\mathrm{id}) = 1$.

Consider $A := (x, L/K) = \bigoplus_{\sigma \in \Gamma} Le_\sigma$, $B := (y, L/K) = \bigoplus_{\sigma \in \Gamma} Lf_\sigma$, and call

$$e'_\sigma := z(\sigma) f_\sigma \in B .$$

Then we find

$$e'_\sigma x = z(\sigma) f_\sigma x = {}^\sigma x z(\sigma) f_\sigma = {}^\sigma x e'_\sigma \quad (x \in L) \text{ as well as}$$

$$e'_\sigma e'_\tau = z(\sigma) f_\sigma z(\tau) f_\tau = z(\sigma)^\sigma z(\tau) y(\sigma, \tau) f_{\sigma\tau} = x(\sigma, \tau) z(\sigma\tau) f_{\sigma\tau} =$$
$$= x(\sigma, \tau) e'_{\sigma\tau} ,$$

hence

$$f: A \longrightarrow B , \quad \sum_{\sigma \in \Gamma} x_\sigma e_\sigma \mapsto \sum_{\sigma \in \Gamma} x_\sigma e'_\sigma$$

is a K-algebra homomorphism and thus an isomorphism because of Lemma 1 in §9. □

Theorem 3. *Let* L/K *be finite Galois,* $\Gamma := \mathrm{Gal}(L/K)$ *and* $x, y \in C^2(\Gamma, L^*)$, *then*

$$(x, L/K) \otimes_K (y, L/K) \sim (xy, L/K) .$$

Note that we write the $\mathbb{Z}$-module $C^2(\Gamma, L^*)$ multiplicatively.

Proof. Set $A := (x, L/K) = \bigoplus_{\sigma \in \Gamma} Le_\sigma$, $B := (y, L/K) = \bigoplus_{\sigma \in \Gamma} Lf_\sigma$, $C :=$
$:= (xy, L/K) = \bigoplus_{\sigma \in \Gamma} Lg_\sigma$, and consider the commutative K-algebra $L \otimes_K L \subseteq$
$\subseteq A \otimes_K B$. Now write $L = K(\theta)$ and consider the (monic) minimal polynomial $f_\theta \in K[T]$ of θ over K. Then

$$f_\Theta(T) = \sum_{i=0}^{n} a_i T^i = \prod_{\sigma \in \Gamma} (T - {}^\sigma\Theta) \in L[T] \quad (a_n = 1, n = |L:K|).$$

Consider the element

$$e := \prod_{id \neq \sigma \in \Gamma} \frac{\Theta \otimes 1 - 1 \otimes {}^\sigma\Theta}{\Theta \otimes 1 - {}^\sigma\Theta \otimes 1} \in L \otimes_K L \subseteq A \otimes_K B,$$

and regard f_Θ as an element in $K \otimes_K K[T]$ rather than in $K[T]$. Then

$$e(\Theta \otimes 1 - 1 \otimes \Theta) = \prod_{\sigma \in \Gamma} \frac{(\Theta \otimes 1 - 1 \otimes {}^\sigma\Theta)}{\text{denominator}} = \frac{f_\Theta(\Theta \otimes 1)}{\text{denominator}} = \frac{f_\Theta(\Theta) \otimes 1}{\text{denominator}} = 0, \text{ hence}$$

$e(\Theta \otimes 1) = (1 \otimes \Theta)e$, and therefore (by induction) $e(\Theta^i \otimes 1) = (1 \otimes \Theta^i)e$, consequently (because of $L = \bigoplus_{i=0}^{n-1} K\Theta^i$)

(2) $\quad e(x \otimes 1) = (1 \otimes x)e \text{ for all } x \in L$.

Now fix $\tau \in \Gamma$ and define

$$e^{(\tau)} := \prod_{id \neq \sigma \in \Gamma} \frac{{}^\tau\Theta \otimes 1 - 1 \otimes {}^{\tau\sigma}\Theta}{{}^\tau\Theta \otimes 1 - {}^{\tau\sigma}\Theta \otimes 1} \in L \otimes_K L \subseteq A \otimes_K B;$$

using (2) $(n-1)$-times and taking into account that $L \otimes_K L$ is commutative gives

(3) $\quad ee^{(\tau)} = e$, in particular (in the case $\tau = id$): $e^2 = e$.

By our construction the numerator of e is a polynomial of degree $n-1$ in $\Theta \otimes 1$ over $K \otimes_K L$ with constant term $\pm 1 \otimes \frac{1}{\Theta} N_{L/K}(\Theta) \neq 0$. Therefore - because of $L \otimes_K L = \bigoplus_{i=0}^{n-1} (\Theta \otimes 1)^i (K \otimes_K L)$ - we get $e \neq 0$, hence

$e \in L \otimes_K L \subseteq A \otimes_K B$ is an idempotent $\neq 0$.

Let us now consider the ring $e(A \otimes_K B)e =: C'$ with unit element e (cf. §3.). Fix $\sigma, \tau \in \Gamma$, then

$$e(e_\sigma \otimes f_\tau)e = e(e_\sigma \otimes f_\tau) \prod_{id \neq \rho \in \Gamma} \frac{\Theta \otimes 1 - 1 \otimes {}^\rho\Theta}{\Theta \otimes 1 - {}^\rho\Theta \otimes 1} = e \prod_{id \neq \rho \in \Gamma} \frac{{}^\sigma\Theta \otimes 1 - 1 \otimes {}^{\tau\rho}\Theta}{{}^\sigma\Theta \otimes 1 - {}^{\sigma\rho}\Theta \otimes 1}(e_\sigma \otimes f_\tau) =$$

$$= ee^{(\sigma)}(e_\sigma \otimes f_\sigma) = e(e_\sigma \otimes f_\sigma) \quad \text{for} \quad \sigma = \tau \text{ (see (2))},$$
$$= 0 \qquad\qquad\qquad\qquad\qquad\qquad \sigma \neq \tau \text{ (see (3))}.$$

Define $g'_\sigma := e(e_\sigma \otimes f_\sigma)e \in C'$; from the last calculation and from (2)/(3) it follows immediately

$$e\Big(\big(\sum_{\sigma \in \Gamma} x_\sigma e_\sigma\big) \otimes \big(\sum_{\tau \in \Gamma} y_\tau f_\tau\big)\Big)e = \sum_{\sigma,\tau \in \Gamma} e(x_\sigma \otimes 1)(1 \otimes y_\tau)(e_\sigma \otimes f_\tau)e =$$

$$= \sum_{\sigma \in \Gamma} (x_\sigma y_\sigma \otimes 1)g'_\sigma \quad (x_\sigma, y_\sigma \in L), \quad g'_\sigma g'_\tau = e(e_\sigma \otimes f_\sigma)e^2(e_\tau \otimes f_\tau)e =$$

$$= e(e_\sigma e_\tau \otimes f_\sigma f_\tau)e = e(x(\sigma,\tau) \otimes y(\sigma,\tau))(e_{\sigma\tau} \otimes f_{\sigma\tau})e =$$

$$= (x(\sigma,\tau)y(\sigma,\tau) \otimes 1)g'_{\sigma\tau} \text{ and } g'_\sigma(x \otimes 1) = e(e_\sigma x \otimes f_\sigma)e = e({}^\sigma x e_\sigma \otimes f_\sigma)e =$$

$$= (^\sigma x \otimes 1)g'_\sigma \quad (x \in L , \sigma \in \Gamma) .$$

Consequently
$$f: C \longrightarrow C' , \sum_{\sigma \in \Gamma} x_\sigma g_\sigma \mapsto \sum_{\sigma \in \Gamma} (x_\sigma \otimes 1)g'_\sigma$$
is a surjective K-algebra homomorphism, hence an isomorphism (cf. Lemma 1 in §9.). Now *(ii)* in Wedderburn's Main Theorem in §3. shows $C' \sim A \otimes_K B$, therefore $C \sim A \otimes_K B$. □

Summarizing all the material hitherto discussed in this paragraph gives the

Crossed Product Theorem. *Let L/K be finite Galois, $\Gamma := \mathrm{Gal}(L/K)$, then the assignment $x \mapsto [x,L/K]$ ($x \in C^2(\Gamma,L^*)$, cf. Theorem 2) induces an isomorphism*
$$\Omega_{L/K}: H^2(\Gamma,L^*) \xrightarrow{\sim} \mathrm{Br}(L/K) .$$

Proof. By Theorem 2/Lemma 1 $\Omega_{L/K}$ is a well-defined map; Theorem 3 says that this map is in fact a homomorphism which is injective thanks to the second part of Theorem 1 (see *(v) ibid.*). Finally $\Omega_{L/K}$ is also surjective; this follows from the first part of Theorem 1 together with Theorem 7 in §9. □

Exercise 1. Let M be a left Γ-module, $\Gamma = \langle\sigma\rangle$ a finite cyclic group; define for $m \in M^\Gamma$ a function $x_{m,\sigma}: \Gamma \times \Gamma \to M$ by
$$x_{m,\sigma}(\sigma^i,\sigma^j) := \begin{smallmatrix}0\\m\end{smallmatrix} \text{ if } i + j \begin{smallmatrix}<\\\geq\end{smallmatrix} n \quad (0 \leq i,j < n ; n = |\Gamma|) .$$
Show that $x_{m,\sigma} \in C^2(\Gamma,M)$ and that the assignment $m \mapsto x_{m,\sigma}$ induces an isomorphism $\Omega_\sigma: H^0(\Gamma,M) \xrightarrow{\sim} H^2(\Gamma,M)$. Moreover, in the situation of Theorem 1 in §10. show

(4) $\qquad [a,L/K,\sigma] = [x_{a,\sigma},L/K]$ in $\mathrm{Br}(K)$.

Exercise 2. Let M,N be left Γ-modules, $x \in C^1(\Gamma,M)$ and $y \in C^1(\Gamma,N)$. Define the *cup product* $x \cup y: \Gamma \times \Gamma \to M \otimes_Z N$ by
$$(x \cup y)(\sigma,\tau) := x(\sigma) \otimes {}^\sigma y(\tau) .$$
Show that $x \cup y \in C^2(\Gamma,M \otimes_Z N)$ and that the assignment $(x,y) \mapsto x \cup y$ induces a Z-bilinear skew symmetric map
$$H^1(\Gamma,M) \times H^1(\Gamma,N) \longrightarrow H^2(\Gamma,M \otimes_Z N) .$$
(Here the "skew symmetric" is to be understood modulo the canonical isomorphism $M \otimes_Z N \simeq N \otimes_Z M$).

Exercise 3. Deduce Noether's Equations (in §6.) from the results of this paragraph (cf. M. Deuring [1935], p.66).

§ 13. THE FORMALISM OF CROSSED PRODUCTS

Let M be a left Γ-module, $\Delta \trianglelefteq \Gamma$ a normal subgroup and $G := \Gamma/\Delta$ the factor group, then M^Δ carries (in a natural way) the structure of a left G-module via ${}^{\bar{\sigma}}m := {}^{\sigma}m$ (here $\sigma \in \bar{\sigma}$, i.e. $\sigma\Delta = \bar{\sigma} \in G$). Moreover, if $x \in C^i(G, M^\Delta)$ ($i = 1,2$) then we define a function χ by
$$\chi(\sigma) := x(\bar{\sigma}) \quad \text{(if } i=1 \text{)} \quad \text{resp.} \quad \chi(\sigma, \tau) := x(\bar{\sigma}, \bar{\tau}) \quad \text{(if } i=2 \text{)}$$
(here $\sigma\Delta = \bar{\sigma}$, $\tau\Delta = \bar{\tau}$; $\bar{\sigma}, \bar{\tau} \in G$),
and it is easily seen that $\chi \in C^i(\Gamma, M)$ ($i = 1,2$) holds. A straightforward calculation gives then:

Lemma 1. *In the situation described above the assignment $x \mapsto \chi$ induces a homomorphism*
$$\inf^i_{\Gamma/\Delta} : H^i(G, M^\Delta) \longrightarrow H^i(\Gamma, M) \quad (i = 1, 2)$$
which is called the "inflation". □

Theorem 1. *Let L/K be finite Galois of degree n, $\Gamma := \text{Gal}(L/K)$, and consider an intermediate field I of degree m over K which is likewise Galois over K. Let $\Delta := \text{Gal}(L/I) \trianglelefteq \Gamma$, hence $s := \frac{n}{m} = |L:I| = |\Delta|$ and $G := \Gamma/\Delta = \text{Gal}(I/K)$. Then, if $x \in C^2(G, I^*)$ is given, we have*
$$(x, L/K) \simeq M_s\big((x, I/K)\big)$$
(χ as in Lemma 1), i.e. the diagram shown commutes. (Here the vertical arrows are the isomorphisms from the Crossed Product Theorem in §12.)

$$\begin{array}{ccc} H^2(G, I^*) & \xrightarrow{\inf^2_{\Gamma/\Delta}} & H^2(\Gamma, L^*) \\ \Omega_{I/K} \downarrow \wr & & \wr \downarrow \Omega_{L/K} \\ \text{Br}(I/K) & \hookrightarrow & \text{Br}(L/K) \end{array}$$

Proof. The proof is almost identical with the proof of Lemma 8 in §10.: every $x \in L$ resp. $\sigma \in \Gamma$ determines a matrix
$$f(x) \in M_s(I) \quad \text{resp.} \quad F(\sigma) \in GL_s(I)$$
such that (note that Γ acts on $M_s(I)$ componentwise)
(1) $\qquad f({}^\sigma x) F(\sigma) = F(\sigma) {}^\sigma f(x) \quad (x \in L, \sigma \in \Gamma)$

and
(2) $\quad\quad\quad F(\sigma\tau) = F(\sigma)^{\sigma} F(\tau) \quad (\sigma,\tau \in \Gamma)$.

Now set $A := (x,L/K) = \bigoplus_{\sigma \in \Gamma} Le_{\sigma}$, $A := (x,I/K) = \bigoplus_{\delta \in G} Ie_{\delta}$ and define

$$f(e_{\sigma}) := F(\sigma)e_{\delta} \quad (\sigma \in \Gamma \,;\, \delta = \sigma\Delta \in G).$$

It follows from (1)

$$f(e_{\sigma})f(x) = F(\sigma)e_{\delta}f(x) = F(\sigma)^{\delta}f(x)e_{\delta} = F(\sigma)^{\sigma}f(x)e_{\delta} =$$
$$= f(^{\sigma}x)F(\sigma)e_{\delta} = f(^{\sigma}x)f(e_{\sigma}) \quad (x \in L \,;\, \sigma \in \Gamma,\, \delta = \sigma\Delta \in G)$$

and from (2) - note $f(x) = x1$ if $x \in I$ -

$$f(e_{\sigma})f(e_{\tau}) = F(\sigma)e_{\delta}F(\tau)e_{t} = F(\sigma)^{\delta}F(\tau)e_{\delta}e_{t} = F(\sigma)^{\sigma}F(\tau)x(\delta,t)e_{\delta t} =$$
$$= F(\sigma\tau)x(\sigma,\tau)e_{\delta t} = f(x(\sigma,\tau))F(\sigma\tau)e_{\delta t} = f(x(\sigma,\tau))f(e_{\sigma\tau})$$
$$(\sigma,\tau \in \Gamma \,;\, \delta = \sigma\Delta,\, t = \tau\Delta \,;\, \delta,t \in G).$$

Consequently a K-algebra homomorphism $f: A \longrightarrow M_s(A)$ is well-defined and even an isomorphism thanks to Lemma 1 in §9. □

A nice application of the above is:

Theorem 2. *Assume* $char(K) \nmid m$ *($m \in \mathbb{N}$ given) and take* $[A] \in {}_m Br(K)$. *Then there exists a finite Galois extension* L/K *with Galois group* Γ *such that* $\mu_m \subset L$ *and*

$$[A] = [y,L/K] \text{ with suitable } y \in C^2(\Gamma,\mu_m) \subseteq C^2(\Gamma,L^*).$$

Proof. Thanks to Theorem 9 in §9. and the Crossed Product Theorem in §12. we can find a finite Galois extension I/K with Galois group G such that $\mu_m \subset L$ and

$$[A] = [x,I/K] \text{ for some } x \in C^2(G,I^*).$$

By assumption we have then (thanks to $m[A] = 0$)

$$x(\delta,t)^m = \frac{z(\delta)^{\delta} z(t)}{z(\delta t)} \quad (\delta, t \in G)$$

for some function $z: G \to I^*$ satisfying $z(id) = 1$. Now select L/K finite Galois such that $I, \mu_m \subset L$ and $\sqrt[m]{z(\delta)} \in L$ for all $\delta \in G$. Setting

$$y(\sigma,\tau) := x(\sigma,\tau)\frac{z(\sigma)^{\sigma} z(\tau)}{z(\sigma\tau)} \text{ with } x \text{ as in Lemma 1 } \textbf{and}$$
$$z(\sigma) := (\sqrt[m]{z(\delta)})^{-1} \quad (\sigma,\tau \in \Gamma \,;\, \delta = \sigma\Delta \in G \,;\, \Delta := Gal(L/I) \trianglelefteq$$
$$\trianglelefteq Gal(L/K) =: \Gamma \text{ and } \Gamma/\Delta = G)$$

we easily find $y(\sigma,\tau)^m = 1$ - hence $y(\sigma,\tau) \in \mu_m$ - and (cf. Theorem 1 and the Crossed Product Theorem in §12.)

$$[A] = [x,I/K] = [x,L/K] = [y,L/K] \quad . \; □$$

Let M be a left Γ-module, then it is in particular a left Δ-module for any subgroup $\Delta \leq \Gamma$; hence the following is obvious:

Lemma 2. *In the situation described above any given* $x \in C^i(\Gamma,M)$ ($i =$
$= 1,2$) *may be viewed (by restricting the variables to* Δ *) as an element
in* $C^i(\Delta,M)$, *and this procedure induces a homomorphism*
$$\mathrm{res}^i_{\Gamma/\Delta} : H^i(\Gamma,M) \longrightarrow H^i(\Delta,M) \quad (i = 1,2)$$
which is called the "restriction". □

Theorem 3. *Let* L/K *be finite Galois,* $\Gamma := \mathrm{Gal}(L/K)$, I *an intermediate field and* $\Delta := \mathrm{Gal}(L/I) \leq \Gamma$. *Then, if* $x \in C^2(\Gamma,L^*)$ *is given and viewed as an element in* $C^2(\Delta,L^*)$ *(see Lemma 2), we have*
$$(x,L/K) \otimes_K I \sim (x,L/I) ,$$
i.e. the diagram shown on the right commutes. (Again the vertical arrows are the isomorphisms from the Crossed Product Theorem in §12.)

$$\begin{array}{ccc} H^2(\Gamma,L^*) & \xrightarrow{\mathrm{res}^2_{\Gamma/\Delta}} & H^2(\Delta,L^*) \\ \Omega_{L/K} \downarrow & & \downarrow \Omega_{L/I} \\ \mathrm{Br}(L/K) & \xrightarrow{r_{I/K}} & \mathrm{Br}(L/I) \end{array}$$

Proof. Write $A := (x,L/K) = \bigoplus_{\sigma \in \Gamma} Le_\sigma$ and compute $Z_A(I)$; consider $a = \sum_{\sigma \in \Gamma} x_\sigma e_\sigma \in A$, then $a \in Z_A(I)$ if and only if we have for all $x \in I$
$$0 = xa - ax = \sum_{\sigma \in \Gamma} (x - {}^\sigma x) x_\sigma e_\sigma = \sum_{\sigma \notin \Delta} (x - {}^\sigma x) x_\sigma e_\sigma$$
which amounts to $x_\sigma = 0$ for all $\sigma \notin \Delta$. Hence we conclude (cf. Corollary 1 in §9.)
$$(x,L/K) \otimes_K I = A \otimes_K I \sim Z_A(I) = \bigoplus_{\delta \in \Delta} Le_\delta = (x,L/I) . \square$$

The next result is a supplement to the previous.

Theorem 4. *Let* L/K *be finite Galois with Galois group* Γ *and let* F/K *be an arbitrary (not necessarily finite) extension such that* $L \cap F = K$. *Then* $L \otimes_K F$ *is a field* $\simeq LF$ *such that* $L \otimes_K F/F$ *is likewise Galois with mutatis mutandis the same Galois group* Γ *(via the correspondence* $\sigma \mapsto \sigma \otimes \mathrm{id}_F$; *here we abbreviate* $\sigma \otimes \mathrm{id}_F$ *by* σ *after identifying* $L \otimes_K F$ *with* LF *). Moreover, we may view any* $x \in C^2(\Gamma,L^*)$ *as an element in* $C^2(\Gamma,(LF)^*)$ *and in this sense we have*
$$(x,L/K) \otimes_K F \simeq (x,LF/F) .$$

Proof. All but the last assertion is well-known from Field Theory. As for the formula concerning the crossed products we note that there is an obvious injection $(x,L/K) \to (x,LF/F)$; F-linear extension thereof (cf. Theorem 3 in §5.) and Lemma 1 in §9. give the required isomorphism. □

We should remark that if in Theorem 4 we drop the assumption

"L ∩ F = K" it would not do any harm in the following sense: consider the intermediate field I := L ∩ F of the extension L/K and use Theorem 3 before using Theorem 4 (with I in place of K). This shows that in *all* cases $(x,L/K) \otimes_K F$ may be computed in terms of crossed products.

Let Γ be a group, $\Delta \trianglelefteq \Gamma$ a normal subgroup and M a left Δ-module. Then, if $x \in C^i(\Delta,M)$ (i = 1,2) and $\sigma \in \Gamma$ are given, we define a function $^\sigma x$ by

$$(^\sigma x)(\delta) := {^\sigma x}(\sigma^{-1}\delta\sigma) \text{ (if i=1) resp. } (^\sigma x)(\delta,\varepsilon) :=$$
$$:= {^\sigma x}(\sigma^{-1}\delta\sigma, \sigma^{-1}\varepsilon\sigma) \text{ (if i=2) } (\delta,\varepsilon \in \Delta),$$

and it is easily seen that $^\sigma x \in C^i(\Delta,M)$ (i = 1,2) holds. A straightforward calculation gives then:

Lemma 3. *In the situation described above the assignment* $x \mapsto {^\sigma x}$ *induces a homomorphism*
$$\mathrm{con}^i_\sigma : H^i(\Delta,M) \longrightarrow H^i(\Delta,M) \quad (i = 1,2)$$
which is called the "conjugation". □

Theorem 5. *Let* L/K *be finite Galois,* $\Gamma := \mathrm{Gal}(L/K)$, *and consider an intermediate field* I *which is likewise Galois over* K. *Let* $\Delta :=$
$:= \mathrm{Gal}(L/I) \trianglelefteq \Gamma$, *hence* $G := \Gamma/\Delta = \mathrm{Gal}(I/K)$. *Then, if* $x \in C^2(\Delta,L^*)$ *and* $\sigma \in \Gamma$ *are given, we have*

$$^\delta(x,L/I) \simeq (^\sigma x, L/I)$$

($\delta = \sigma\Delta \in G$), i.e. the diagram shown commutes. (Here the vertical arrows are the isomorphisms from the Crossed Product Theorem *in §12.)*

$$\begin{array}{ccc} H^2(\Delta,L^*) & \xrightarrow{\mathrm{con}^2_\sigma} & H^2(\Delta,L^*) \\ \downarrow \Omega_{L/I} & & \downarrow \Omega_{L/I} \\ \mathrm{Br}(L/I) & \xrightarrow{[A] \mapsto [^\delta A]} & \mathrm{Br}(L/I) \end{array}$$

Proof. Thanks to Lemma 2 in §8. and Lemma 1 in §9. it suffices to establish a ring homomorphism
$$f: (^\sigma x, L/I) \longrightarrow (x, L/I)$$
such that $f(x) = \delta^{-1}(x) = \sigma^{-1}(x)$ for all $x \in I$. For this purpose let $A := (x,L/I) = \bigoplus_{\delta \in \Delta} L e_\delta$, $B := (^\sigma x, L/I) = \bigoplus_{\delta \in \Delta} L f_\delta$ and define f via the assignments
$$x \mapsto {^{\sigma^{-1}}x} \quad (x \in L) \text{ and } f_\delta \mapsto e_{\sigma^{-1}\delta\sigma} \quad (\delta \in \Delta).$$

Thanks to
$$e_{\sigma^{-1}\delta\sigma} {^{\sigma^{-1}}x} = {^{\sigma^{-1}}(^\delta x)} e_{\sigma^{-1}\delta\sigma} \quad (x \in L, \delta \in \Delta)$$
and
$$e_{\sigma^{-1}\delta\sigma} e_{\sigma^{-1}\varepsilon\sigma} = x(\sigma^{-1}\delta\sigma, \sigma^{-1}\varepsilon\sigma) e_{\sigma^{-1}\delta\varepsilon\sigma} = {^{\sigma^{-1}}((^\sigma x)(\delta,\varepsilon))} e_{\sigma^{-1}\delta\varepsilon\sigma}$$

($\delta, \varepsilon \in \Delta$) we see that our definition of f is feasible. □

Of course we may interpret Lemma 3 such that conjugation endows $H^i(\Delta, M)$ ($i = 1, 2$) with the structure of a left Γ-module. In this context one can show that $H^i(\Delta, M)$ is a trivial Δ-module, hence a Γ/Δ-module (cf. Exercise 1); in the special situation of Theorem 5 the latter is clear !

Moreover, if we combine Theorems 1/3/5 with (6) and Theorem 4 in §9. we see that there is an exact sequence

(3) $\qquad 1 \longrightarrow H^2(G, I^*) \xrightarrow[\inf^2_{\Gamma/\Delta}]{} H^2(\Gamma, L^*) \xrightarrow[\text{res}^2_{\Gamma/\Delta}]{} H^2(\Delta, L^*)^G$

and a short exact sequence

(4) $\qquad 1 \longrightarrow H^2(G, I^*) \longrightarrow H^2(\Gamma, L^*) \longrightarrow H^2(\Delta, L^*)^G \longrightarrow 1$
if $G = \Gamma/\Delta$ is cyclic.

Both sequences are valid (under certain conditions) in the general situation (see Exercise 1).

Now let M be a left Γ-module, $\Delta \leq \Gamma$ a subgroup of finite index n and R a system of representatives for the cosets of Γ modulo Δ, i.e. $\Gamma = \bigcup_{\rho \in R} \rho\Delta$, $|R| = |\Gamma:\Delta| = n$. Given $\rho \in R$ and $\sigma \in \Gamma$, there are (uniquely determined) elements

(5) $\qquad ^\sigma\rho \in R$ and $\delta(\sigma, \rho) \in \Delta$ such that $\sigma\rho = {}^\sigma\rho \, \delta(\sigma, \rho)$

satisfying

(6) $\qquad ^{\sigma\tau}\rho = {}^\sigma({}^\tau\rho)$, $^1\rho = \rho$; $\delta(\sigma\tau, \rho) = \delta(\sigma, {}^\tau\rho)\delta(\tau, \rho)$, $\delta(1, \rho) = 1$

(cf. (9) and (10) in §8.). Now take any $x \in C^i(\Delta, M)$ ($i = 1, 2$) and define a function $c_R x$ by

$(c_R x)(\sigma) := \sum_{\rho \in R} {}^\rho x(\delta(\sigma, \rho))$ (if $i=1$) resp. $(c_R x)(\sigma, \tau) :=$

$:= \sum_{\rho \in R} {}^{\sigma\tau}\rho \, x(\delta(\sigma, {}^\tau\rho), \delta(\tau, \rho))$ (if $i=2$) (here $\sigma, \tau \in \Gamma$).

A somewhat lengthy (but entirely straightforward) calculation shows that $c_R x \in C^i(\Gamma, M)$ holds. Moreover, an even longer calculation shows (the details are left to the reader, cf. Exercise 3)

Lemma 4. *In the situation described above the assignment* $x \mapsto c_R x$ *induces a homomorphism*

$$\text{cor}^i_{\Gamma/\Delta} : H^i(\Delta, M) \longrightarrow H^i(\Gamma, M) \quad (i = 1, 2)$$

which is independent of the choice of R and called the "corestriction". Moreover ($i = 1, 2$)

$$\text{cor}^i_{\Gamma/\Delta} \text{res}^i_{\Gamma/\Delta} = |\Gamma:\Delta| \text{id} \quad (\text{id} := \text{identity on } C^i(\Gamma, M))$$

and

$$\operatorname{res}^i_{\Gamma/\Delta} \operatorname{cor}^i_{\Gamma/\Delta} = N_G$$

if $\Delta \leq \Gamma$ *is a normal subgroup,* $G := \Gamma/\Delta$. □

Theorem 6 . *Let* L/K *be finite Galois,* $\Gamma := \operatorname{Gal}(L/K)$, I *an intermediate field and* $\Delta := \operatorname{Gal}(L/I) \leq \Gamma$. *Then, if* $x \in C^2(\Delta, L^*)$ *is given, we have*

$$c_{I/K}(x, L/I) \sim (c_R x, L/K)$$

($c_R x$ *as in* Lemma 4), *i.e. the diagram shown commutes. (Here the vertical arrows are the isomorphisms from the* Crossed Product Theorem *in §12.)*

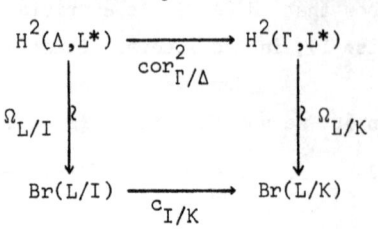

Proof. We omit the (rather technical) proof of this theorem (see Exercise 3) since we shall nowhere make use of this result in these lectures. □

Exercise 1 . Let Γ be a group, $\Delta \trianglelefteq \Gamma$ a normal subgroup, $G := \Gamma/\Delta$ the factor group and N a left Δ-module. Show that $H^i(\Delta, N)$ is a left G-module via conjugation (i = 1,2). Now let M be a left Γ-module; prove the *inflation restriction sequences*, i.e. show that the following sequences are well-defined and exact:

(7) $\qquad 0 \longrightarrow H^1(G, M^\Delta) \xrightarrow{\inf^1} H^1(\Gamma, M) \xrightarrow{\operatorname{res}^1} H^1(\Delta, M)^G$

and

(8) $\qquad 0 \longrightarrow H^2(G, M^\Delta) \xrightarrow{\inf^2} H^2(\Gamma, M) \xrightarrow{\operatorname{res}^2} H^2(\Delta, M)^G$

if $H^1(\Delta, M) = \{0\}$.

Note that Noether's Equations in §6. and (8) imply (3) !

Exercise 2 . Establish commutation rules between inflation/restriction/conjugation/corestriction (cf. Lemmas 1/2/3/4) and the isomorphism $\Omega_{./.}$ from Exercise 1 in §12. in the case where Γ is a finite *cyclic* group.

Exercise 3 . Prove Lemma 4 and Theorem 6 . (For a proof of Theorem 6 see Theorem 11 in C. Riehm [1970]; however, note that there a description of the isomorphism $\Omega_{./.}$ in the Crossed Product Theorem (in §12.) is used which is not exactly the one we introduced (cf. J.-P. Serre [1962, p.166]).)

§ 14 . QUATERNION ALGEBRAS

In §11. we introduced *quaternion algebras*
$\left(\frac{a,b}{K}\right)$ ($a,b \in K^*$) resp. $\left(\frac{a,b}{K}\right]$ ($a \in K^*$, $b \in K$)
if char(K) $\neq$ 2 resp. char(K) = 2 (cf. Definitions 2 resp. 6 in §11.) as special cases of power norm residue algebras, hence we may use all results of §11. in order to describe the behaviour of quaternion algebras. It is worth rewriting these results in the special notation of quaternion algebras; here we shall use the special notation

(1) $\quad\quad\left(\frac{a,b}{K}\right)\right]$ if the respective rule applies regardless of char(K).

We start with the (almost obvious)

Theorem 1 . *A* K-*algebra* A *is a quaternion algebra if and only if it is a central simple* K-*algebra of reduced degree* 2 *(i.e.* $|A:K| = 2^2 = 4$ *)*.

Proof. The "only if" is trivial; "if" : because of Theorem 1*(f)* in §9. we have *either* $A \simeq M_2(K)$ - in which case A obviously is a quaternion algebra - *or* A is a skew field (of index 2). In the latter case A has a separable quadratic - and thus over K *cyclic* - subfield (cf. Köthe's Theorem in §9.), hence A is a cyclic algebra thanks to Lemma 2 in §10. Now our assertion is clear from Lemma 2/Corollary 3 resp. Theorem 5 in §11. if char(K) $\neq$ 2 resp. = 2 . □

Theorems 1/4 in §11. read as follows:

Theorem 2 . *All* K-*algebras* A *generated by two elements* a,b *which are subject to the relations*
$\quad\quad a := a^2 \in K^*$, $b := b^2 \in K^*$, $ab = -ba$
are isomorphic to $\left(\frac{a,b}{K}\right)$ *if* char(K) $\neq$ 2 . □

Theorem 3 . *All* K-*algebras* A *generated by two elements* a,b *which are subject to the relations*

$a := a^2 \in K^*$, $b := b^2 + b \in K$, $ab = (b+1)a$ are isomorphic to $\left(\frac{a,b}{K}\right]$ if char(K) = 2 (i.e. -1 = 1). □

Theorems 2/3 imply
$$D_a(L/K) \simeq \left(\frac{a,b}{K}\right)]$$
with suitable b depending on L where the left hand side is according to (2) in §1. Moreover, one sees
$$H \simeq \left(\frac{-1,-1}{R}\right) \ .$$

Now if we rewrite Lemmas 4/8/14 & Corollaries 3/4/6 in §11. we find the following (cf. the convention introduced in (1))

Rules for quaternion algebras .

(2) $\left(\frac{a,b}{K}\right) \simeq M_2(K)$ if and only if the equation $a = x^2 - y^2 b$ is soluble over K ;

(3) $\left(\frac{a,b}{K}\right] \simeq M_2(K)$ if and only if the equation $a = x^2 + xy + y^2 b$ is soluble over K ;

(4) $\left(\frac{a,b}{K}\right) \simeq \left(\frac{b,a}{K}\right)$;

(5) $\left(\frac{a',b}{K}\right)] \otimes_K \left(\frac{a'',b}{K}\right)] \sim \left(\frac{a'a'',b}{K}\right)]$;

(6) $\left(\frac{a,b'}{K}\right) \otimes_K \left(\frac{a,b''}{K}\right) \sim \left(\frac{a,b'b''}{K}\right)$;

(7) $\left(\frac{a,b'}{K}\right] \otimes_K \left(\frac{a,b''}{K}\right] \sim \left(\frac{a,b'+b''}{K}\right]$;

(8) $\left(\frac{a,b}{K}\right)] \otimes_K L \simeq \left(\frac{a,b}{L}\right)]$. □

Another interesting result is (use Lemmas 1/12 in §11. & Lemma 3(β) in §9.)

Theorem 4 . *Assume* $|K(\sqrt{a}):K| = 2$ *and let* $K(\sqrt{a})$ *be a splitting field of the quaternion algebra* $A' := \left(\frac{a',b'}{K}\right)]$, *then* $A' \simeq \left(\frac{a,b}{K}\right)]$ *for suitable* b . □

For the next result recall the definition introduced in the course of Exercise 1 in §9.; then use Corollary 3 in §9., Lemma 2 in §10. and Theorem 5 in §11. This gives

Theorem 5 . *Assume* $|K(\frac{1}{\wp}b):K| = 2$ *and let* $K(\frac{1}{\wp}b)$ *be a splitting field of the quaternion algebra* $A' := \left(\frac{a',b'}{K}\right]$, *then* $A' \simeq \left(\frac{a,b}{K}\right]$ *for suitable* a . □

The following results on quaternion algebras seem to be not so well-known.

Theorem 6. *Let* A', A'' *be quaternion algebras over* K; *then* $A' \otimes_K A''$ *is not a skew field if and only if* A' *and* A'' *have a common splitting field which is separable quadratic over* K.

Proof. cf. Exercise 1. □

With the aid of (4) and Theorems 4/5 one can restate the preceding theorem in the following way:

Corollary 1. *Assume* $\left(\frac{a',b'}{K}\right) \otimes_K \left(\frac{a'',b''}{K}\right) \sim \left(\frac{a,b}{K}\right)$, *then one can find* c', c'', d *such that* $\left(\frac{a',b'}{K}\right) \simeq \left(\frac{c',d}{K}\right)$ *and* $\left(\frac{a'',b''}{K}\right) \simeq \left(\frac{c'',d}{K}\right)$. □

Interesting enough one can show that (under the same assumptions as in Corollary 1) in the case "char(K) = 2" it is *not always possible* to find suitable c, d', d'' such that $\left(\frac{a',b'}{K}\right) \simeq \left(\frac{c,d'}{K}\right)$ and $\left(\frac{a'',b''}{K}\right) \simeq \left(\frac{c,d''}{K}\right)$ (cf. (4.26) on p.134 in R. Baeza [1978]).

Theorem 7. *Assume* $\left(\frac{a',b'}{K}\right) \simeq \left(\frac{a'',b''}{K}\right)$, *then there exists an element* d *such that* $\left(\frac{a',b'}{K}\right) \simeq \left(\frac{a',d}{K}\right) \simeq \left(\frac{a'',d}{K}\right) \simeq \left(\frac{a'',b''}{K}\right)$.

Proof. cf. Exercise 2. □

Exercise 1. Prove Theorem 6 (see A.A. Albert [1972] and P. Draxl [1975]).

Exercise 2. Prove Theorem 7 (see J. Tate [1976,p.267] if char(K) $\neq$ 2).

Exercise 3. Recall that a field is called *formally real* if -1 cannot be expressed as a sum of squares. Now call a field *Pythagorean* if it is formally real and every sum of squares therein is a square.(Note that a formally real field must have characteristic 0.) Show that a field K is Pythagorean if and only if $\left(\frac{-1,-1}{K}\right)$ is a skew field such that any of its maximal commutative subfields is K-isomorphic to $K(\sqrt{-1})$ (see B. Fein & M. Schacher [1976]).

We close this paragraph with a remark: the reader should know that the theory of quaternion algebras is closely linked to the theory of *quadratic forms* via the *Clifford algebras*. Standard references for that are for instance O.T. O'Meara [1963] and T.Y. Lam [1973].

§ 15. p-Algebras

In this paragraph we investigate the p-primary component $Br(K)_p$ of the Brauer group $Br(K)$ of a field K with $p := char(K) \neq 0$.

Definition 1. *Let* $p := char(K) \neq 0$, *then any central simple K-algebra A such that* $|A:K|$ = p-power *is called a "p-algebra over K".*

From the various results of §9. the following is clear:

Lemma 1. *Let A be a p-algebra over K, then* $i(A)$ = p-power *and* $[A] \in Br(K)_p$. *Conversely, if* $i(A)$ = p-power *or (equivalently)* $[A] \in Br(K)_p$, *then the skew field component D of A is a p-algebra over K.* □

Clearly the algebras $(a,b;p,K>$ from §11. are p-algebras and it is clear that they have the purely inseparable splitting field $K(\sqrt[p]{a})$ over K. The last observation is just an example to the following general

Theorem 1. *Let A be a p-algebra over K, then* $[A] \in Br(I/K)$ *for some purely inseparable extension* I/K *such that* $I^{o(A)} \subseteq K$.

Corollary 1. *Let K be a perfect field of characteristic* $p \neq 0$, *then* $Br(K)_p = \{0\}$, *i.e.* $p \nmid i(A)$ *for all* $[A] \in Br(K)$. □

Proof. Put $o(A) = p^e$ with suitable $e \in \mathbb{N}$ (in case $e = 0$ there is nothing to show). Now write (cf. §12. and Theorem 9 in §9.)

$$[A] = [x, L/K] \text{ for some finite Galois } L/K \text{ with } \Gamma := Gal(L/K)$$
$$\text{and } x \in C^2(\Gamma, L^*).$$

The Crossed Product Theorem in §12. implies then

$$x(\sigma,\tau)^{p^e} = \frac{z(\sigma)^\sigma z(\tau)}{z(\sigma\tau)} \qquad (\sigma,\tau \in \Gamma)$$

for some function $z: \Gamma \to L^*$ satisfying $z(id) = 1$. Now consider the field

$$M := L(\{\sqrt[p^e]{{}^\tau z(\sigma)} | \sigma,\tau \in \Gamma\}), \text{ i.e. } M/L \text{ is purely inseparable}.$$

If $\gamma \in \Gamma$ one can extend γ to an automorphism of M by setting

$$\gamma(\sqrt[p^e]{\tau z(\sigma)}) := \sqrt[p^e]{\gamma\tau z(\sigma)} \quad,$$

hence Γ may be viewed as a group of K-automorphisms of the field M. Now set

$$I := M^\Gamma \quad, \text{ then } M/I \text{ is Galois with group } \Gamma \ .$$

Moreover, $x \in I \subseteq M$ implies $x^{p^e} \in L$ (by definition of M) and therefore $\gamma(x^{p^e}) = x^{p^e}$ (by definition of I), hence $x^{p^e} \in K$. Concluding we have $I \cap L = K$ and $M = IL \simeq I \otimes_K L$, i.e. we may use Theorem 4 in §13.: if we set

$$u(\sigma) := \sqrt[p^e]{z(\sigma)} \quad (\sigma \in \Gamma)$$

we find

$$x(\sigma,\tau) = \frac{u(\sigma)^\sigma u(\tau)}{u(\sigma\tau)} \quad, \text{ hence } x \in B^2(\Gamma,M^*)$$

and therefore (cf. also the Crossed Product Theorem in §12.)

$$r_{I/K}[A] = r_{I/K}[x,L/K] = [x,LI/I] = [x,M/I] = 0 \quad . \quad \square$$

Before we proceed we need two results from Field Theory which appear not to be so widely known. The first of it is a special case of deep results due to E. Witt [1936]:

Lemma 2. *Assume* $p := \text{char}(K) \neq 0$ *and let* L/K *be a finite cyclic extension of degree* p^f *with generating automorphism* τ *. Then there exists a cyclic extension* N/K *of degree* p^{f+1} *with generating automorphism* σ *such that* $K \subseteq L \subseteq N$ *and* $\sigma|_L = \tau$ *.*

Proof. Choose an element $a \in L$ such that $\text{Tr}_{L/K}(a) = 1$ (cf. (1) in §6.); now use the additive endomorphism $\wp$ of L^+ (see §11.) which commutes with $\text{Tr}_{L/K}$ as well as with τ (this is clear from Field Theory), hence

$$\text{Tr}_{L/K}(\wp a) = \wp(\text{Tr}_{L/K}(a)) = \wp 1 = 0 \quad .$$

Therefore Corollary 1 in §6. implies

$$\wp a = {}^\tau b - b \text{ for suitable } b \in L \ .$$

We claim $b \notin \wp L$. Indeed, if we had $b = \wp c$ for some $c \in L$ we would get $\wp a = {}^\tau b - b = {}^\tau(\wp c) - \wp c = \wp({}^\tau c - c)$, hence (cf. the exact sequence (12) in §11.) $a = {}^\tau c - c + x$ with some $x \in F_p \subseteq K$, and the latter would then lead to the contradiction $1 = \text{Tr}_{L/K}(a) = \text{Tr}_{L/K}(x) = p^f x = 0$. Now define (cf. Exercise 1 in §9.)

$$N := L(\tfrac{1}{\wp}b) = L(d) \text{ where } \wp d = b \text{ and } N/L \text{ is cyclic of degree } p \text{ with generating automorphism } \gamma: d \mapsto d+1 \ .$$

Furthermore, extend τ to an automorphism σ of N via $d \mapsto d + a$.

It follows $\sigma^{p^f} d = \ldots = d + \sum_{i=0}^{p^f} \tau^i a = d + Tr_{L/K}(a) = d + 1 = {}^\gamma d$, hence σ has order p^{f+1} in $Aut(N)$. Therefore N/K is the required field extension. □

An immediate application of Lemma 2 is

Lemma 3. *Let* $p := char(K) \neq 0$ *and* $[A] = [a,L/K,\tau] \in Br(K)$, *then there exists a* $[B] \in Br(K)$ *such that* $[A] = p[B]$ *in* $Br(K)$.

Proof. Take $[B] := [a,N/K,\sigma] \in Br(K)$ (in the sense of Lemma 2) and use Lemmas 4/8 in §10. □

The second result from Field Theory which we shall need is the following

Lemma 4. *Let* I/K *be a finite and purely inseparable extension and* L'/I *finite separable (resp. Galois); then there exists a separable (resp. Galois) extension* L/K *such that* $L \otimes_K I \simeq LI = L'$. *Moreover, in the Galois case* $\Gamma' := Gal(L'/I)$ *and* $\Gamma := Gal(L/K)$ *can be identified via the correspondences* $\sigma' \mapsto \sigma'|_L$ *(* $\sigma' \in \Gamma'$ *) and* $\sigma \mapsto \sigma \otimes id_I$ *(* $\sigma \in \Gamma$ *).* □

Now we are fully prepared for

Theorem 2. *Let* $p := char(K) \neq 0$, $|K(\sqrt[p^e]{a}):K| = p^e$ *and* $[A] \in Br(K(\sqrt[p^e]{a})/K)$, *then*

$$[A] = [a,L/K,\sigma] \text{ in } Br(K)$$

for some cyclic extension L/K *of degree* p^e.

Proof. We proceed by induction on e; the case $e = 1$ has been settled in Lemma 12 in §11. (see also its proof). Now assume $e > 1$, set

$$F := K(\sqrt[p^e]{a}) \text{ and } I := K(\sqrt[p]{a}) \text{, hence } |F:I| = p^{e-1} \text{ and}$$
$$r_{I/K}[A] \in Br(F/I) .$$

Consequently the induction hypothesis implies

$$r_{I/K}[A] = [\sqrt[p]{a}, L_0/I, \sigma_0] \text{ for some cyclic } L_0/I \text{ of degree } p^{e-1}.$$

Now let us use Lemmas 2/4 simultaneously: this gives a cyclic extension L_1/K of degree p^e such that L_1I/I is also cyclic of degree p^e with generating automorphism σ_1 where $I \subseteq L_0 \subseteq L_1I$ and $\sigma_1|_{L_0} = \sigma_0$.

Lemmas 7/8 in §10. yield then

$$r_{I/K}[A] = [\sqrt[p]{a}, L_0/I, \sigma_0] = [a, L_1I/I, \sigma_1] = r_{I/K}[a, L_1/K, \sigma_1] .$$

The latter implies

$$[A] - [a, L_1/K, \sigma_1] \in Br(K(\sqrt[p]{a})/K) \text{ and } |K(\sqrt[p]{a}):K| = p ,$$

hence (see the case $e = 1$) for some cyclic L_2/K of degree p:

(1) $\qquad [A] = [a, L_1/K, \sigma_1] + [a, L_2/K, \sigma_2]$.

From Field Theory we know that *either* $L_1 \cap L_2 = K$ *or* $L_2 \subseteq L_1$. In the *first* case the assertion follows then immediately from (1) because of Lemma 9 and Theorem 1 in §10., whereas in the *second* we can find $t \in \mathbb{N}$ coprime with p such that $\sigma_1\big|_{L_2} = \sigma_2^t$. Now, if $s(1+tp^{e-1}) \equiv 1 \mod p^e$ (note that $1+tp^{e-1}$ and p^e are coprime), Lemmas 3/8 in §10. imply

$$[a, L_2/K, \sigma_2] = [a^t, L_2/K, \sigma_2^t] = [a^{tp^{e-1}}, L_1/K, \sigma_1],$$

and therefore we may conclude from (1) thanks to Lemmas 3/4 in §10.

$$[A] = (1+tp^{e-1})[a, L_1/K, \sigma_1] = (1+tp^{e-1})[a^s, L_1/K, \sigma_1^s] =$$
$$= [a, L/K, \sigma] \text{ where } L := L_1 \text{ and } \sigma := \sigma_1 \text{ . } \square$$

Theorem 3. *Let* $p := \mathrm{char}(K) \neq 0$, $I := K(\sqrt[p^{e_1}]{a_1}, \ldots, \sqrt[p^{e_r}]{a_r})$ *and assume* $|I:K| = \prod_{i=1}^{r} p^{e_i}$; *then, if* $[A] \in \mathrm{Br}(I/K)$,

$$[A] = \sum_{i=1}^{r} [a_i, L_i/K, \sigma_i]$$

for suitable cyclic extensions L_i/K.

Proof. We proceed by induction on r; the case $r = 1$ has been settled in the preceding theorem. Now suppose $r > 1$ and set

$$F := K(\sqrt[p^{e_2}]{a_2}, \ldots, \sqrt[p^{e_r}]{a_r}) \subseteq I \text{ , hence } I = F(\sqrt[p^{e_1}]{a_1}),$$

$$r_{F/K}[A] \in \mathrm{Br}(I/F) \text{ and } |I:F| = p^{e_1}.$$

It follows (use Theorem 2 and Lemma 4 together with Lemma 7 in §10.)

$$r_{F/K}[A] = [a_1, L_1 F/F, \sigma_1] = r_{F/I}[a_1, L_1/K, \sigma_1]$$

which amounts to

$$[A] - [a_1, L_1/K, \sigma_1] \in \mathrm{Br}(F/K) \text{ where } |F:K| = \prod_{i=2}^{r} p^{e_i}$$

and therefore to our assertion (by construction of F) thanks to the induction hypothesis. $\square$

Now let $[A] \in \mathrm{Br}(K)_p$ be given and select a minimal purely inseparable extension I/K such that $[A] \in \mathrm{Br}(I/K)$ and $I^{o(A)} \subseteq K$ (such a field I exists thanks to Theorem 1). Then I is of the form described in Theorem 3 and we get the important

Theorem 4. *Assume* $p := \mathrm{char}(K) \neq 0$ *and take* $[A] \in \mathrm{Br}(K)_p$, $o(A) = p^e$. *then there exist elements* $a_1, \ldots, a_r \in K^*$, $0 \leq e_1 \leq \ldots \leq e_r = e$ *and cyclic extensions* $L_1/K, \ldots, L_r/K$ *such that*

$$[A] = \sum_{i=1}^{r} [a_i, L_i/K, \sigma_i] \text{ . } \square$$

More old-fashionedly one would state Theorem 4 in the form

Corollary 2. *Every p-algebra over K is similar to a tensor product of cyclic algebras.* □

Moreover, if in Theorem 4 we denote by $L := L_1..L_r$ the composite field of the cyclic fields L_i over K, then L/K is abelian, hence

Corollary 3. *If $p := \mathrm{char}(K) \neq 0$, then every $[A] \in \mathrm{Br}(K)_p$ has an abelian splitting field over K.* □

Witt's Theorem. *Let $p := \mathrm{char}(K) \neq 0$, then $\mathrm{Br}(K)$ is p-divisible.*

Proof. Only the p-primary component $\mathrm{Br}(K)_p$ matters since $\mathrm{Br}(K)$ is a torsion group (cf. Theorem 10 in §9.), however, the p-divisibility of $\mathrm{Br}(K)_p$ is clear from Lemma 3 and Theorem 4. □

Another important result is (compare it with Theorem 4 in §9.)

Hochschild's Theorem. *Let L/K be a finite purely inseparable extension, then the sequence*
$$0 \longrightarrow \mathrm{Br}(L/K) \subseteq \longrightarrow \mathrm{Br}(K) \longrightarrow \mathrm{Br}(L) \longrightarrow 0$$
is exact.

Proof. Let $[A] \in \mathrm{Br}(L)$ be given and choose e such that $L^{p^e} \subseteq K$. Using Witt's Theorem repeatedly we can find $[B] \in \mathrm{Br}(L)$ such that $[A] = p^e[B]$. Now take F/L finite Galois with Galois group Γ such that
$$[B] = [x, F/L] \text{ for some } x \in C^2(\Gamma, F^*)$$
(cf. Theorem 9 in §9. and the Crossed Product Theorem in §12.). Thanks to Lemma 4 we have $F = NL$ with some finite Galois extension N/K with Galois group Γ (via restriction of the action of the K-automorphisms of F to N) and such that $F^{p^e} \subseteq N$, hence
$$x^{p^e} \in C^2(\Gamma, N^*),$$
and therefore (see the Crossed Product Theorem in §12. and Theorem 4 in §13.) the equation
$$[A] = p^e[B] = [x^{p^e}, NL/L] = r_{L/K}[x^{p^e}, N/K] \quad . \quad □$$

The theory of p-algebras culminates in

Albert's Main Theorem. *Assume $p := \mathrm{char}(K) \neq 0$ and take $[A] \in \mathrm{Br}(K)_p$, then there exist a cyclic extension L/K and $a \in K^*$ such that*
$$[A] = [a, L/K, \sigma],$$
i.e. A has a cyclic splitting field over K.

Proof. see A.A. Albert[1939,p.109] or Exercises 1/2 . □

Although Albert's Main Theorem is much stronger than Theorem 4 and Corollary 3 the latter two results are sufficient for most applications. Moreover one should point out that a p-algebra A over K is *not necessarily isomorphic* to a cyclic algebra (see O. Teichmüller [1936,pp.386]). Finally we remark that the reader may find different (and most interesting) approaches to the theory of p-algebras in E. Witt [1937] and G. Hochschild [1955].

Exercise 1 . Assume $p := \text{char}(K) \neq 0$, set $I := K(\sqrt[p^e]{a})$ ($a \in K^*$) and let L/I be finite separable. Show $I = K(N_{L/I}(b))$ for some $b \in L^*$.

Exercise 2 . Use Exercise 1 for a proof of the following result (which is clearly sufficient to prove Albert's Main Theorem): if A,B are cyclic p-algebras over K , then $A \otimes_K B$ is isomorphic to a cyclic p-algebra over K .

Exercise 3 . Let p be a prime $\neq \text{char}(K)$ and assume $\langle \zeta \rangle = \mu_p \subset K$. Moreover, let L/K be a finite cyclic extension of degree p^f with generating automorphism τ such that $\zeta \in N_{L/K}(L^*)$. Show that there exists a cyclic extension N/K of degree p^{f+1} with generating automorphism σ such that $K \subseteq L \subseteq N$ and $\sigma|_L = \tau$.(Hint. Modify the proof of Lemma 2 .)

§ 16 . SKEW FIELDS WITH INVOLUTION

Let A be a ring (with $1 \neq 0$), then we call as usual a Z-module automorphism θ of A a ring *antiautomorphism* of A if $\theta(x)\theta(y) = \theta(yx)$ for all $x,y \in A$. In this context the following is obvious:

(1) If θ is a ring antiautomorphism of A , then θ defines an isomorphism $A \simeq A^{op}$.

(2) If θ is a ring antiautomorphism of A , then the same is true for θ^{-1} .

(3) If θ, Ω are ring antiautomorphisms of A , then $\theta\Omega$ is a ring automorphism of A .

(4) If θ is a ring antiautomorphism of A , then, for any $a \in A^*$ $\theta_a : A \to A$, $x \mapsto a\theta(x)a^{-1}$ is again a ring antiautomorphism of A .

(5) Any ring antiautomorphism θ of A can be extended to a ring antiautomorphism of $M_n(A)$ via $\theta(x_{ij}) := (\theta(x_{ji}))$.

Definition 1 . *Let A be a ring (with $1 \neq 0$) and I a ring antiautomorphism of A such that $I^2 = id_A$, then I is called an "involution of A " and we write ^Ix in place of $I(x)$ ($x \in A$). Moreover, we denote by $S_I(A) := \{ x \in A \mid {}^Ix = x \}$ the Z-module of " I-symmetric elements" in A .*

In what follows let K be a commutative field and A a K-algebra satisfying $Z(A) = K$ (e.g. a central simple K-algebra). Then, if I is an involution of A it must preserve the centre (argue as if I were an automorphism; cf. (12) in §7.), hence *either* $I|_K = id_K$ (in case $K \subseteq S_I(A)$) *or* $I|_K \neq id_K$ (in case $k := K \cap S_I(A) \neq K$) .

Definition 2 . *Let A be a K-algebra with $K = Z(A)$. An involution I of A is called of the "first kind" (resp. "second kind") if*
$$k := K \cap S_I(A) = K \text{ (resp. } \neq K \text{).}$$

Let us investigate the case of central simple K-algebras; first, we observe that then (4) has a converse, namely:

Lemma 1. *If Θ, Ω are ring antiautomorphisms of the central simple K-algebra A such that $\Theta|_K = \Omega|_K$, then $\Omega = \Theta_a$ (in the sense of (4)) for suitable $a \in A^*$ (which is unique modulo K^*).*

Proof. Thanks to (2) and (3) $\Theta^{-1}\Omega$ is a ring automorphism of A and even a K-algebra automorphism (because of $\Theta|_K = \Omega|_K$), hence

$$\Theta^{-1}\Omega(x) = txt^{-1} \quad \text{for some (modulo } K^* \text{ unique) } t \in A^*$$

by virtue of the Skolem-Noether Theorem in §7., i.e.

$$\Omega(x) = a\Theta(x)a^{-1} = \Theta_a(x) \quad (x \in A) \text{ where } a := \Theta(t) \in A^* . \quad \square$$

Second, we may restrict our attention to K-skew fields thanks to

Lemma 2. *Let D be a K-skew field and $A := M_r(D)$. Then, if Θ is a ring antiautomorphism of A there exists an $a \in A^*$ such that Θ_a arises from a ring antiautomorphism of D as described in (5).*

Proof. Take the matrices $e_{ij} \in M_r(D) = A$ (cf. §2.) and consider the elements $f_{ij} := \Theta(e_{ji}) \in A$ which satisfy the identities

$$f_{ij}f_{rs} = \Theta(e_{ji})\Theta(e_{sr}) = \Theta(e_{sr}e_{ji}) = \begin{matrix} 0 & \text{if} & j \neq r \\ \Theta(e_{si}) = f_{is} & & j = r \end{matrix} ,$$

hence

$$\bigoplus_{i=1}^{r} \bigoplus_{j=1}^{r} K e_{ij} \simeq \bigoplus_{i=1}^{r} \bigoplus_{j=1}^{r} K f_{ij} \quad \text{as K-subalgebras of A} ,$$

and therefore Corollary 2 in §7. implies

(6) $\qquad e_{ij} = af_{ij}a^{-1} = \Theta_a(e_{ji})$ for a suitable $a \in A^*$.

Now take $d \in D$; it follows

$$\Theta_a(d)e_{ij} = \Theta_a(d)\Theta_a(e_{ji}) = \Theta_a(e_{ji}d) = \Theta_a(de_{ji}) =$$
$$= \Theta_a(e_{ji})\Theta_a(d) = e_{ij}\Theta_a(d) ,$$

hence

(7) $\qquad \Theta_a(d) \in Z_{M_r(D)}(M_r(K)) = D$ for all $d \in D$.

But (6) and (7) amount clearly to the fact that Θ_a arises from a ring antiautomorphism of D as described in (5). $\square$

Corollary 1. *Let A be a central simple K-algebra. The fact whether or not A admits a ring antiautomorphism (e.g. an involution) depends only on its class [A] in the Brauer group Br(K).* $\square$

Now we are prepared for the following result due to A.A. Albert:

Theorem 1. *Let* [A] $\in$ Br(K), *then* A *admits an involution of the first kind if and only if* 2[A] = 0 *(i.e.* [A] $\in {}_2$Br(K) *).*

Proof. The "only if" is clear since the isomorphism $A \simeq A^{op}$ from (1) is then a K-algebra isomorphism, hence $[A] = [A^{op}] = -[A]$. For the proof of the "if" we note that by virtue of Corollary 1, Theorem 9 in §9. and the Crossed Product Theorem in §12. we may suppose

$$A = (x, L/K) = \bigoplus_{\sigma \in \Gamma} Le_\sigma \quad \text{where } L/K \text{ is finite Galois with}$$

Galois group Γ and $x \in C^2(\Gamma, L^*)$ such that $x(\sigma,\tau)^2 =$

$$= \frac{z(\sigma)^\sigma z(\tau)}{z(\sigma\tau)} \quad (\sigma,\tau \in \Gamma \, ; \, z: \Gamma \to L^* \text{ such that } z(\text{id}) = 1),$$

hence

(8) $\qquad x(\sigma,\tau)^{-1} z(\sigma)^\sigma z(\tau) = z(\sigma\tau) x(\sigma,\tau)$.

Now define

$$I: A \longrightarrow A \, , \, \sum_{\sigma \in \Gamma} x_\sigma e_\sigma \mapsto \sum_{\sigma \in \Gamma} e_\sigma^{-1} z(\sigma) x_\sigma \; ;$$

clearly I is a Z-module endomorphism of A such that (cf. (8))

$$I(xe_\sigma y e_\tau) = I(x^\sigma y x(\sigma,\tau) e_{\sigma\tau}) = e_{\sigma\tau}^{-1} z(\sigma\tau) x(\sigma,\tau) x^\sigma y =$$

$$= e_{\sigma\tau}^{-1} x(\sigma,\tau)^{-1} z(\sigma)^\sigma z(\tau) x^\sigma y = (e_\sigma e_\tau)^{-1} z(\sigma)^\sigma z(\tau) x^\sigma y =$$

$$= (e_\tau^{-1} z(\tau) y)(e_\sigma^{-1} z(\sigma) x) = I(y e_\tau) I(x e_\sigma) \quad (x, y \in L \, ; \, \sigma,\tau \in \Gamma)$$

and consequently

$$I^2(e_\sigma) = I(e_\sigma^{-1} z(\sigma)) = I(z(\sigma)) I(e_\sigma^{-1}) = z(\sigma) I(e_\sigma)^{-1} =$$

$$= z(\sigma)(e_\sigma^{-1} z(\sigma))^{-1} = e_\sigma \quad (\sigma \in \Gamma) \, ,$$

hence I is the involution we were looking for. □

From Theorem 1 it is clear that any tensor product of quaternion algebras admits an involution of the first kind. Just recently it has been proved (see A.C. Меркурев [1981]; §17.) that the converse is *mutatis mutandis* also true, namely:

(9) If a central simple algebra admits an involution of the first kind, then it is similar to a tensor product of quaternion algebras.

For further information on this cf. J.-P. Tignol [1981] and the various references there.

Now let us focus our attention on the case of central simple K-algebras admitting involutions of the second kind. We begin with a slightly more general situation: if A is a K-algebra such that K = Z(A) and if A admits an involution I of the second kind, then we have an automorphism $\sigma := I|_K \neq \mathrm{id}_K$ of K such that $\sigma^2 = \mathrm{id}_K$, i.e. the non-trivial k-automorphism σ of the separable quadratic field extension K/k ($k := K \cap S_I(A)$) is extended to our ring antiautomorphism I of A. This suggests the following

Definition 3. *Let K/k be a separable quadratic field extension with Galois group $\Gamma := \{\sigma, \mathrm{id}\}$ and A a K-algebra such that $K = Z(A)$, then a ring antiautomorphism Θ of A satisfying $\Theta|_K = \sigma$ is called a " K/k-antiautomorphism of A ". Moreover, if a K/k-antiautomorphism is an involution, we call it a " K/k-involution".*

From now on we present the extremely elegant arguments due to W. Scharlau [1975].

Scharlau's Lemma. *Let K/k be separable quadratic, $\Gamma := \mathrm{Gal}(K/k) = \{\sigma, \mathrm{id}\}$ and Θ a K/k-antiautomorphism of the central simple K-algebra A, then there exists an element $b \in A^*$ such that*

(10) $\Theta^2(x) = bxb^{-1}$ ($x \in A$).

Moreover, if we take any $b \in A^*$ such that (10) is fulfilled, then

(11) $b\Theta(b) = \Theta(b)b \in k^*$

holds and the class of $b\Theta(b)$ in $k^*/N_{K/k}(K^*) = H^0(\Gamma, K^*)$ depends only on the class $[A]$ in the Brauer group $\mathrm{Br}(K)$. Finally, if we replace Θ by Θ_a and b by $b_a := a\Theta(a)^{-1}b$, then

$$\Theta_a^2(x) = b_a x b_a^{-1} \text{ and } b_a \Theta_a(b_a) = b\Theta(b).$$

Proof. (10) is clear from (3) together with the Skolem-Noether Theorem in §7. because Θ^2 must be a K-algebra automorphism thanks to $\sigma^2 = \mathrm{id}_K$. Now we deduce from (10) for all $x \in A$ the equations

$$b\Theta(x)b^{-1} = \Theta^2(\Theta(x)) = \Theta(\Theta^2(x)) = \Theta(bxb^{-1}) = \Theta(b)^{-1}\Theta(x)\Theta(b),$$

hence $\Theta(b)b \in Z(A)^* = K^*$ and consequently

$$(b\Theta(b))^2 = b(\Theta(b)b)\Theta(b) = (\Theta(b)b)(b\Theta(b)), \text{ i.e. } b\Theta(b) = \Theta(b)b,$$

as well as

$$\sigma(\Theta(b)b) = \Theta(\Theta(b)b) = \Theta(b)\Theta^2(b) = \Theta(b)bbb^{-1} = \Theta(b)b,$$

hence (11). Now we note that b in (10) is unique modulo K^*, so, if we replace b by bc ($c \in K^*$) we get

$$bc\Theta(bc) = bc\Theta(c)\Theta(b) = b\Theta(b)c^\sigma c = b\Theta(b)N_{K/k}(c).$$

Now replace Θ by Θ_a etc.; it follows

$$\Theta_a^2(x) = a\Theta(a\Theta(x)a^{-1})a^{-1} = a\Theta(a)^{-1}bxb^{-1}\Theta(a)a^{-1} = b_a x b_a^{-1}$$

and

$$b_a \Theta_a(b_a) = a\Theta(a)^{-1} b a \Theta(a\Theta(a)^{-1} b) a^{-1} =$$
$$= a\Theta(a)^{-1} b a (\Theta(b)b) a^{-1} b^{-1} \Theta(a) a^{-1} = \Theta(b)b = b\Theta(b) .$$

Finally assume $A = M_r(D)$ for some K-skew field D. By the preceding arguments and thanks to Lemma 2 we may assume

$$\Theta(D) = D \quad \text{and} \quad \Theta(d_{ij}) = (\Theta(d_{ji})) .$$

By virtue of

$$\Theta^2(x_{ij}) = \Theta(\Theta(x_{ji})) = (\Theta^2(x_{ij})) \quad (x_{ij} \in D)$$

we may then choose $b \in D^* \subseteq A^*$ in (10); this completes our proof. □

From Scharlau's Lemma one obtains

Scharlau's Criterion. *Let K/k be separable quadratic and $[A] \in Br(K)$, then A admits a K/k-involution if and only if it admits a K/k-antiautomorphism Θ such that $b\Theta(b) \in N_{K/k}(K^*)$ (here $\Theta^2(x) = bxb^{-1}$).*

Proof. The "only if" is trivial, for if I is a K/k-involution we may take $\Theta = I$ and $b = 1$. For the proof of the "if" we use an argument due to R. Scharlau (W. Scharlau's original argument was slightly more complicated): thanks to Lemma 2 and Scharlau's Lemma we may restrict our attention to the case where A is a K-skew field. Now assume $b\Theta(b) =$
$= N_{K/k}(c) = c\Theta(c)$ for some $c \in K^*$, then - if $d := bc^{-1}$ -
$$d\Theta(d) = 1 \quad \text{and} \quad \Theta^2(d) = d ,$$
and we may consider the commutative subfield $L := K(d)$ of our skew field A. Obviously $\Theta(L) = L$ and $\Theta^2|_L = id_L$. Now *either* $\Theta|_L = id_L$, hence $d^2 = 1$ and therefore $b \in K^*$ which means that $I := \Theta$ is our involution, *or* $\Theta|_L \neq id_L$. In the latter case Hilbert's "Satz 90" in §6. yields an element $a \in L \subseteq A$ such that $b = \Theta(a)a^{-1}$. Now take $I := \Theta_a$ which gives
$$I^2(x) = \Theta_a^2(x) = a\Theta(a\Theta(x)a^{-1})a^{-1} = a\Theta(a)^{-1} bxb^{-1}\Theta(a)a^{-1} = x . \quad \square$$

K/k-involutions (i.e. involutions of the second kind) are in relationship with the corestriction from §8. This comes from

Scharlau's Theorem. *Let K/k be separable quadratic, $\Gamma := Gal(K/k) =$ $= \{\sigma, id\}$ and $[A] \in Br(K)$. Then the following conditions are equivalent:*
(i) $\quad A$ *admits a K/k-antiautomorphism Θ ;*
(ii) $\quad {}^\sigma A \simeq A^{op}$ *as K-algebras;*
(iii) $\quad r_{K/k}(c_{K/k}[A]) = 0$ *in $Br(K)$;*
(iv) $\quad A$ *admits a K/k-antiautomorphism Θ such that we have*
$$c_{K/k}[A] = [b\Theta(b), K/k, \sigma] \quad \text{in } Br(k) \quad (\text{here } \Theta^2(x) = bxb^{-1}).$$

Proof. *(iv)* implies *(i)* trivially and *(i)* implies *(ii)* because the isomorphism from (1) is in fact a K-algebra isomorphism $^\sigma A \simeq A^{op}$ given by Θ (cf. Lemma 1 in §8.; note $\sigma = \sigma^{-1}$). *(ii)* implies *(iii)* thanks to (cf. Theorem 5 in §9.)

$$r_{K/k}(c_{K/k}[A]) = N_\Gamma[A] = [A] + [^\sigma A] = [A] + [A^{op}] = 0 .$$

Now, since the same calculation shows that *(iii)* implies *(i)* it suffices to deduce *(iv)* from *(i)*, *(ii)*, *(iii)*.

Indeed, from §8. we know that $C := c_{K/k}(A) = (A \otimes_K {}^\sigma A) \subseteq A \otimes_K {}^\sigma A$ is a k-subalgebra; moreover, if Θ is the K/k-antiautomorphism from *(i)* we can find a K-algebra isomorphism (cf. *(ii)* and Definition 1 in §7.)

$$f: A \otimes_K {}^\sigma A \xrightarrow{\sim} \text{End}_K(A) \quad \text{such that} \quad f(a \otimes b)(x) = ax\Theta(b)(\ x \in A \ ).$$

In particular, we may view A as a left C-module (via f and the embedding of C), and therefore it makes sense to consider the k-algebra $\text{End}_C(A)$. Since *(iii)* implies $[C] = c_{K/k}[A] \in \text{Br}(K/k)$ we get $i(C)|2$ (cf. Corollary 4 in §9.), hence (in what follows now cf. §3.)

$C \sim D$ for the k-skew field $D \simeq \text{End}_C(\hbar)$ where $\hbar$ is a minimal right ideal of C.

On the other hand A is a left C-module (see above), hence a right C^{op}-module, but $C \simeq C^{op}$ thanks to $i(C)|2$, so we may view A likewise as a right C-module, and in this sense we have (see again §3.)

$$(12) \qquad A \simeq \bigoplus_{i=1}^{m} \hbar \quad \text{and consequently} \quad \text{End}_C(A) \simeq M_m(D) .$$

Now set $n := |A:K|$, hence $|C:k| = n^2$ (see Lemma 7 in §8.). Then we get *either* $C \simeq M_n(k)$ (if $i(C) = 1$) and consequently $\dim_k(\hbar) = n$, $mn = |A:k| = 2n$ which amounts to $\text{End}_C(A) \simeq M_2(k)$, *or* $C \simeq M_{n/2}(D)$ (if $i(C) = 2$) and $|D:k| = 4$, hence $\dim_k(\hbar) = 2n$, $m2n = |A:k| = 2n$ and thus $\text{End}_C(A) \simeq D$. In any case we have (after summarizing; cf. (12))

$$(13) \qquad c_{K/k}[A] = [\text{End}_C(A)] \quad \text{in} \quad \text{Br}(k) \quad \text{and} \quad |\text{End}_C(A):k| = 4 .$$

From (13) we conclude with the aid of Lemma 1 in §9. that, in order to complete the proof of *(iv)* it suffices to establish a k-algebra homomorphism (here we view A as a left C-module via f ; see above)

$$g: (b\Theta(b), K/k, \sigma) \longrightarrow \text{End}_C(A) .$$

For this purpose write $(b\Theta(b), K/k, \sigma) = K \oplus Ke$ where $e^2 = b\Theta(b)$ and $ey = \Theta(y)e = {}^\sigma ye$ ($y \in K$) and define g via

$$g(y)(x) := yx \quad (\ y \in K \ ) \quad \text{and}$$
$$g(e)(x) := \Theta(b)\Theta(x) \qquad \text{for all } x \in A .$$

Now a lengthy but entirely straightforward calculation shows that both $g(e), g(y)$ are left C-module endomorphisms of A such that ($y \in K$)
$$g(e)^2 = g(b(b)) \quad \text{and} \quad g(e)g(y) = g(\theta(y))g(e) = g(^\sigma y)g(e) \; . \; \square$$

We close this paragraph by combining Scharlau's Theorem with Scharlau's Criterion which yields (cf. Theorem 1 in §10.)

Riehm's Theorem . *Let* $[A] \in Br(K)$ *, then* A *admits an involution of the second kind if and only if there exists a separable quadratic subfield* k *of* K *such that* $c_{K/k}[A] = 0$ *in* $Br(k)$ *.(The involution is then a K/k-involution.)* $\square$

For further results on involutions cf. W. Scharlau [1981] and Ch.X in A.A. Albert [1939] .

Exercise 1 . Let K/k be separable quadratic and F/k cyclic such that $K \cap F = k$. Set $L := KF \simeq K \otimes_k F$, then L/K is cyclic with generating automorphism (say) γ . Show that the cyclic algebra $(a, L/K, \gamma)$ admits a K/k-involution if and only if $N_{K/k}(a) = N_{F/k}(b)$ for some $b \in F^*$.

Exercise 2 . Let K/k be separable quadratic. Show that a quaternion algebra A over K admits a K/k-involution if and only if $A \simeq B \otimes_k K$ for some quaternion algebra B over k .

§ 17. BRAUER GROUPS AND K_2-THEORY OF FIELDS

In §11. (cf. Definitions 3/4 and Theorem 3 *ibid.*) we have already seen that in case $\mu_n \subset K$ (which implies $\mathrm{char}(K) \nmid n$) the assignment $\{a,b\} \mapsto [a,b;n,K,\zeta]$ induces the *abstract norm residue homomorphism*

(1) $\qquad R_{n,K}: K_2(K)/nK_2(K) \longrightarrow {}_n\mathrm{Br}(K) \otimes_{\mathbb{Z}} \mu_n$

which is independent of the choice of the primitive n-th root of unity ζ. Now recently A.S. Merkur'ev and A.A. Suslin have presented a proof of the following spectacular result:

Merkur'ev-Suslin Theorem . *Let p be a prime $\neq \mathrm{char}(K)$ and assume that every finite field extension of K has p-power degree over K (this implies $\mu_p \subset K$), then $R_{p,K}$ is an isomorphism.* □

It is impossible to give (even a sketch of) the proof within the scope of a book like ours since numerous deep and intrinsic arguments from Algebraic Geometry (e.g. Brauer-Severi varieties), Algebraic K-Theory and Class Field Theory are involved. So we can refer only to the original papers А.С. Меркурев & А.А. Суслин [a],[b].

Later we shall restate the Merkur'ev-Suslin Theorem in a more general setting, however, before we do this let us study the impact of the Merkur'ev-Suslin Theorem on the theory of Brauer groups; we claim that it implies the following (widely conjectured)

Theorem 1 . *Assume $\mu_n \subset K$ and take $[A] \in {}_n\mathrm{Br}(K)$, then*

$$[A] = \sum_{j=1}^{d} [a_j, b_j; n, K, \zeta] \quad \text{for suitable} \quad a_j, b_j \in K^*, \ \langle \zeta \rangle = \mu_n$$

and $d = d(A) \in \mathbb{N}$,

hence A has the abelian splitting field $K(\sqrt[n]{a_1}, \ldots, \sqrt[n]{a_d})$ over K.

Note that Theorem 1 implies Rosset's Theorem in §11., however, *without* an effectively calculable bound for $d(A)$.

Proof. Thanks to Lemma 7 in §11. it suffices to assume $n = p^f$ a p-power.

Let us begin with the case "f=1" ; then - using Galois Theory of infinite extensions (cf. e.g. Ch.I/IV in S.S. Shatz [1972]) - we may consider the fixed field L of a p-Sylow subgroup of the (profinite) Galois group $Gal(\overline{K}/K)$ of a separable closure $\overline{K}$ of K ; obviously L is such that one can apply the Merkur'ev-Suslin Theorem: from the *surjectivity* of $R_{p,L}$ we see that our claim holds over L . Now L can be viewed as a union of fields L_ι such that the extensions L_ι/K are finite of degree coprime with p , hence - after using Corollary 13 in §9. together with the Rosset-Tate Theorem in §11. possibly infinitely many times - our assertion is true even over K if $f = 1$. Suppose $f > 1$; we proceed by induction on f . Assume $\langle \zeta \rangle = \mu_{p^f} \subset K$ and take $[A] \in {}_{p^f}Br(K)$; then $p^{f-1}[A] \in {}_pBr(K)$, hence (see above and use Lemma 6 in §11.)

$$p^{f-1}[A] = \sum_{i=1}^{m} [a_i,b_i;p,K,\zeta^{p^{f-1}}] = \sum_{i=1}^{m} p^{f-1}[a_i,b_i;p^f,K,\zeta]$$

which amounts to

$$[A] - \sum_{i=1}^{m} [a_i,b_i;p^f,K,\zeta] \in {}_{p^{f-1}}Br(K) .$$

Now the induction hypothesis yields (use again Lemma 6 in §11.)

$$[A] - \sum_{i=1}^{m} [a_i,b_i;p^f,K,\zeta] = \sum_{i=m+1}^{d} [a_i,b_i;p^{f-1},K,\zeta^p] =$$
$$= \sum_{i=m+1}^{d} [a_i,b_i^p;p^f,K,\zeta] . \quad \square$$

In precisely the same way as in the case of Rosset's Theorem in §11. we find that Theorem 1 (together with Corollary 3 in §15.) implies

Corollary 1 . *Any central simple K-algebra has a metabelian (and a fortiori soluble) splitting field over* K . $\square$

Some people believe that "metabelian" could be improved to "abelian", however, no reasonable idea for a proof seems to be known. For this and similar questions cf. R.L. Snider [1979].

Now let us make a few additional remarks on the homomorphism $R_{n,K}$ in (1). If one introduces the (profinite) Galois group $\Gamma := Gal(\overline{K}/K)$ of a separable closure $\overline{K}$ of K , then it is possible to interpret the results from §§12/13. in such a way that we even have isomorphisms

$$H^2(\Gamma,\overline{K}^*) \xrightarrow{\sim} Br(K) \quad \text{and} \quad H^2(\Gamma,\mu_n) \xrightarrow{\sim} {}_nBr(K) \quad (char(K) \nmid n)$$

where the H^2 is to be understood as built up from *continuous* 2-cocycles (the profinite group Γ is a topological group) only (one calls this the *continuous cohomology*). Now, if $\mu_n \subset K$, then μ_n is a trivial Γ-module, hence the Z-isomorphism $\mu_n \otimes_Z \mu_n \simeq \mu_n$ (cf. Example 1 in §4.) is even a Γ-isomorphism and we have (see above)

(2) $\qquad H^2(\Gamma, \mu_n \otimes_Z \mu_n) \simeq H^2(\Gamma, \mu_n) \otimes_Z \mu_n \simeq {}_n Br(K) \otimes_Z \mu_n$.

In the general situation (i.e. without the assumption $\mu_n \subset K$) one can still consider $H^2(\Gamma, \mu_n \otimes_Z \mu_n)$, and J. Tate [1976] has established a homomorphism

(3) $\qquad R'_{n,K}: K_2(K)/nK_2(K) \longrightarrow H^2(\Gamma, \mu_n \otimes_Z \mu_n)$ if $char(K) \nmid n$

which coincides with the one from (1) modulo the isomorphisms from (2) in the case where $\mu_n \subset K$ (cf. Theorem (3.1) *ibid.*; for a motivation cf. Exercise 2).

Now, using transfer arguments (formally similar to the arguments from our proof of Theorem 1) one can show (cf. A.C. Меркурев & A.A. Суслин [a],[b]) rather easily that the Merkur'ev-Suslin Theorem can be amplified to the

General Merkur'ev-Suslin Theorem . *The homomorphism from* (3) *(and hence the homomorphism from* (1)*) is an isomorphism.* □

Exercise 1 . Show that ${}_3Br(K)$ is generated by the classes of cyclic algebras. (Hint. Use Theorem 1 together with Theorem 6 in §11.)

Exercise 2 . Assume $\langle \zeta \rangle = \mu_n \subset K$ and L/K finite Galois, $\Gamma := Gal(L/K)$. Show that for any $c \in K^* \cap (L^*)^n$ - i.e. $c = c^n$ for some $c \in L^*$ - $x_c(\sigma) := {}^\sigma c c^{-1}$ defines an element $x_c \in Hom(\Gamma, \mu_n) = C^1(\Gamma, \mu_n)$ which is independent of the choice of c. Now identify μ_n with $\mu_n \otimes_Z \mu_n$ (both are trivial Γ-modules) via $\zeta^i \mapsto \zeta^i \otimes \zeta$ and consider (cf. Exercise 2 in §12.)
$$x_a \cup x_b \in C^2(\Gamma, \mu_n \otimes_Z \mu_n) = C^2(\Gamma, \mu_n) \quad (a,b \in K^* \cap (L^*)^n) .$$

Now show
$$[a,b;n,K,\zeta] = [x_a \cup x_b, L/K] \text{ in } Br(K) \text{ if } |K(\sqrt[n]{a}):K| = n .$$

(Hint. Use Lemma 2 in §11., Exercise 1 in §12. and Theorem 2 in §13.)

§ 18 . A SURVEY OF SOME FURTHER RESULTS

There are many aspects of Brauer groups which we cannot discuss here for various reasons.

First, the relationship with Galois Cohomology (cf. §§12/13. where we established an isomorphism $Br(L/K) \simeq H^2(Gal(L/K),L^*) =: H^2(L/K)$ for Galois extensions L/K) leads to further results if one makes use of general Cohomology Theory: then some of our results appear as special cases of rather general constructions (such as Theorem 4 in §9. and the exact sequence (4) in §13. which are both easy consequences of the Hochschild-Serre spectral sequence). A good reference for this point of view is A. Babakhanian [1972] or E. Weiss [1969].

Second, the relationship with cohomology (see above) may be generalized in the following way: one may establish an isomorphism $Br(L/K) \simeq H^2(L/K)$ even when L/K is not Galois but separable. Then, of course, $H^2(L/K)$ has to be given a new meaning: it is no more a Galois Cohomology group $H^2(Gal(L/K),L^*)$ but an Adamson Cohomology group. Here we refer the reader to the original paper I.T. Adamson [1954].

Third, by introducing even more general cohomology groups - the Amitsur Cohomology groups - one can obtain for instance all our results on $Br(K)$ without making use of Köthe's Theorem (as we do frequently). The main advantage of this method, however, is based on the fact that all this works to a wide extend over a commutative ring R rather than a field K ; in these notes we disregard the notion of the Brauer group over a ring although this theory is today highly developed. The reader should consult for an introduction first F. DeMeyer & E. Ingraham [1971] and then M.-A. Knus & M. Ojanguren [1974].

Fourth, there is a method - called the method of *Polynomial Identity* algebras (*PI*-algebras) - which has been originally developed by S.A. Amitsur for the purpose of furnishing his famous examples of K-skew fields D which are not crossed products (i.e. which contain no maximal commutative subfield which is Galois over K). A selection of references is: the book C. Procesi [1973] and N. Jacobson [1975].

Fifth, much is known in the number theoretic cases, i.e. if K is a *local* field (= a finite extension of either some p-adic field Q_p or $F_q((T))$) or a *global* field (= a finite extension of either Q or $F_q(T)$). It turns out that the theory of Brauer groups over such fields is closely related to Class Field Theory. Here we want to mention two nice results for illustration: the first result is classical (from the 1930's) and very deep in the global case:

(1) *If K is local or global, then every K-skew field D is a cyclic algebra such that* i(D) = o(D) . (cf. M. Deuring [1935, pp.118] or A.A. Albert [1939,pp.149])

The second one has been established only recently and depends on the (also recently established) classification of finite simple groups:

(2) *Let* $K \subset L$ *be global fields such that* $K \neq L$ *, then* Br(L/K) *is infinite.*(see Corollary 4 in B. Fein *et al.* [1981] and more generally B. Fein & M. Schacher [1982])

Basic references (for readers with background in Algebraic Number Theory) are for example: J.-P. Serre [1962], J.W.S. Cassels & A. Fröhlich [1967], A. Weil [1967] and I. Reiner [1975].

For more information on the points mentioned so far in this paragraph see the various articles in the proceedings D. Zelinsky [1976] and M. Kervaire & M. Ojanguren [1981].

Sixth, one may link the theory of (infinite dimensional) skew fields with Functional Analysis in the following way; the well-known Mazur-Gelfand Theorem (cf. e.g. K. Yoshida [1965]) states:

(3) *Every skew field D which is at the same time a normed algebra over* C *is isomorphic to* C .

It is not difficult to deduce from (3) the following generalization of Frobenius' Theorem in §10.:

(4) *Every skew field D which is at the same time a normed algebra over* R *is isomorphic to* R *or* C *or* H .

An immediate consequence of (4) is "Ostrowski's Second Theorem" from Valuation Theory:

(5) *Every skew field D which admits a complete archimedian valuation is isomorphic to* R *or* C *or* H .

Finally, there are many results on $Br(K)$ available in the case where K is a *valued* field (in the general sense of W. Krull); cf. for instance O.F.G. Schilling [1950], W. Scharlau [1969] and P. Draxl [a].

Exercise 1 . Deduce (4) from (3) .

PART III . REDUCED K_1-THEORY OF SKEW FIELDS

The history of reduced K_1-Theory begins with Y. Matsushima & T. Nakayama [1943] who proved (in modern notation) for a K-skew field D

(1) $\quad SK_1(D) = \{1\}$ in case K is a p-adic field .

Later Wang Shianghaw [1950] established a proof of

(2) $\quad SK_1(D) = \{1\}$ in case K is an algebraic number field .

In his proof he used the Grunwald Theorem (on the existence of cyclic number field extensions with prescribed local behaviour) from Class Field Theory. On this occasion, incidentally, Wang Shianghaw discovered the (famous) gap in Grunwald's original proof and suggested a new version of the above mentioned theorem which is nowadays called the Grunwald-Wang Theorem (see e.g. E. Artin & J. Tate [1968]).

It was because of the results (1) and (2) that

(3) $\quad SK_1(D) = \{1\}$

was widely believed to hold *in general*.

Finally V.P. Platonov (cf. В.П. Платонов [1975]) proved that (3) *is false in general* ! In §24. we explain this by introducing an *ad hoc* counter example (which, in order to be understood, requires hardly more knowledge of skew fields than contained in §1. of these lectures). Before this, however, we develop a theory of (the functorial properties of) K_1 and SK_1 over a skew field (see §22/23.) which includes proofs of the *algebraic* part of the classical results (1) and (2) (some *number theoretic* facts needed for the proof can only be stated without proof, for otherwise we would have to go far beyond the scope of these lectures). For a proper treatment of §§22/23. it is necessary to discuss certain results about Dieudonné determinants (of matrices over skew fields). This is done in §20. using a Bruhat normal form introduced in §19. (this is not exactly the customary way to introduce Dieudonné determinants). §21. contains a proof of the simplicity of the projective special linear group over a skew field; this, incidentally has nothing to do with K_1 or SK_1. Finally in §25. we sketch reduced unitary K_1-Theory which connects §16. (Part II) with §§ 22/23/24. and makes use of §21.

§ 19. THE BRUHAT NORMAL FORM

Let R be a ring with $1 \neq 0$; consider the full matrix ring $M_n(R)$ and call (as usual) $GL_n(R) := M_n(R)^*$ its (multiplicative) group of invertible elements. In this context we introduced in §2. certain special matrices, namely (cf. §2. for the exact definition)

$$E_{ij}(t) \quad (\ t \in R\ ;\ i \neq j\ ;\ 1 \leq i,j \leq n\ )\ ,$$
$$D_i(u) \quad (\ u \in R^*\ ;\ 1 \leq i \leq n\ ) \quad \text{and}$$
$$P(\pi) \quad (\ \pi \in S_n\ )$$

with properties

(1) $\quad E_{ij}(t)E_{ij}(t') = E_{ij}(t+t')\ ,\ E_{ij}(0) = 1\ ,\ E_{ij}(t)^{-1} = E_{ij}(-t)\ ,$

(2) $\quad E_{ij}(t)E_{rs}(t') = E_{rs}(t')E_{ij}(t) \quad (\ j \neq r \neq s \neq i \neq j\ )\ ,$

(3) $\quad E_{ij}(tt') = E_{ir}(t)^{-1}E_{rj}(t')^{-1}E_{ir}(t)E_{rj}(t') \quad (\ r \neq i \neq j \neq r\ )\ ,$

(4) $\quad D_i(u)D_i(u') = D_i(uu')\ ,\ D_i(1) = 1\ ,\ D_i(u)^{-1} = D_i(u^{-1})\ ,$

(5) $\quad D_i(u)D_j(u') = D_j(u')D_i(u) \quad (\ i \neq j\ )\ ,$

(6) $\quad P(\pi)P(\pi') = P(\pi\pi')\ ,\ P(\text{id}) = 1\ ,\ P(\pi)^{-1} = P(\pi^{-1}) = P(\pi)^t$

and (cf. Examples 1/2 in §2.)

(7) transforming from A to $\begin{pmatrix} 0 & 1 \\ 1 & 0 \end{pmatrix} A \begin{pmatrix} 0 & 1 \\ 1 & 0 \end{pmatrix}$ amounts to rotating the matrix A by 180 degrees

as well as (more generally than (7))

(8) in the (i,j)-th position of the matrix $P(\pi)^{-1}AP(\pi)$ one finds the element from the $(\pi(i),\pi(j))$-th position of A.

In this context we want to remind the reader of Lemma 1 in §2. which is responsible for most of the rules listed above. For our own convenience we state this lemma again (this time, however, we do so in three steps):

(9) transforming from A to $E_{ij}(t)A$ (resp. $AE_{ji}(t)$) amounts to adding the left(resp. right) t-multiple of the j-th row (resp. column) to the i-th row(resp. column),

(10) transforming from A to $D_i(u)A$ (resp. $AD_i(u)$) amounts to multiplying the i-th row(resp. column) from the left(resp. right) by u

and

(11) transforming from A to $P(\pi)A$ (resp. $AP(\pi)^{-1}$) amounts to moving the i-th row(resp. column) into the position of the $\pi(i)$-th row(resp. column).

Now we are fully prepared for the important

Theorem 1. *Let D be a skew field and $A \in M_n(D)$ such that either its rows are left linearly independent over D or $AB = 1$ for some $B \in$ $\in M_n(D)$; then there exists a decomposition*

(12) $A = TUP(\pi)V$ where $T = \begin{pmatrix} 1 & 0 \\ & \ddots & \\ * & & 1 \end{pmatrix}, U = \begin{pmatrix} u_1 & & 0 \\ & \ddots & \\ 0 & & u_n \end{pmatrix}$ ($u_i \neq 0$),

$V = \begin{pmatrix} 1 & & * \\ & \ddots & \\ 0 & & 1 \end{pmatrix}$ and $\pi \in S_n$ such that $P(\pi)VP(\pi)^{-1} = \begin{pmatrix} 1 & & 0 \\ & \ddots & \\ 0 & & 1 \end{pmatrix}$

are uniquely determined by A.

Definition 1. *The decomposition of an $A \in M_n(D)$ according to (12) is called the "strict Bruhat normal form of A".*

Proof. Let us first show that the T, U, V, π in (12) are uniquely determined: indeed let

$TUP(\pi)V = T'U'P(\pi')V'$ be two such decompositions,

then

(13) $U^{-1}T^{-1}T'U' = P(\pi)VV'^{-1}P(\pi')^{-1}$

where the left-hand side of (13) is a lower triangular matrix. On the other hand we find

(14) $VV'^{-1} = 1 + N$ where N is an upper triangular matrix with 0's in the main diagonal (hence a nilpotent matrix).

Obviously there is no position (i,j) where both matrices on the right-hand side of (14) have an entry $\neq 0$; thanks to (8) this remains true for the right-hand side of (cf. (13) and (14))

(15) $P(\pi)VV'^{-1}P(\pi')^{-1} = P(\pi)P(\pi')^{-1} + P(\pi)NP(\pi')^{-1}$.

On the other hand we know that the matrix $P(\pi)P(\pi')^{-1} = P(\pi\pi'^{-1})$ has in each row and each column exactly one 1 and n-1 0's ; therefore this matrix must be the unit matrix 1 - and this amounts to $\pi' = \pi$ -, **since** otherwise (by what we have observed concerning the right-hand side of (15)) the left-hand side of (13) could not be a lower triangular matrix. Hence the right-hand side of (13) can be rewritten

$$= P(\pi)VP(\pi)^{-1}(P(\pi)V'P(\pi)^{-1})^{-1}$$

and is a lower triangular matrix which is of the form $\begin{pmatrix} 1 & & \square \\ & \ddots & \\ 0 & & 1 \end{pmatrix}$ (cf. the conditions on V and π in (12)).

Therefore both sides of (13) are equal to 1, and this implies (note that $\pi' = \pi$ is already known) $V' = V$ as well as $T'U' = TU$. Clearly the latter implies $T' = T$ and $U' = U$, and this completes the proof of the uniqueness of the strict Bruhat normal form.

Now let us turn to the existence proof: according to our assumptions on A we know that its first row has at least one entry $\neq 0$; let $(1,\rho(1))$ be the left-most position in the first row with such an entry $\neq 0$; then n-1 row operations of the type described in (9) will transform our original matrix A to the new matrix A' as shown:

$$A' := \begin{pmatrix} 0-0 \; u_1 & * \\ * & 0 \\ & \vdots & * \\ & 0 \end{pmatrix} \begin{matrix} \rho(1)\text{st} \\ \text{column} \end{matrix}$$

More precisely, $A' = \prod_{i=n}^{2} E_{i1}(-t_{i1})A$ for some elements t_{i1}. Now consider the second row of A': if the rows of A are left linearly independent over D, then the same is true for the rows of A' since left multiplication with a matrix of the type $E_{ij}(t)$ does not affect this property (see (9)); if - on the other hand - $AB = 1$ for some matrix B, then $A'B' = 1$ where $B' := B \prod_{i=2}^{n} E_{i1}(t_{i1})$ (see (1)). Hence we may conclude that the second row of A' has at least one entry $\neq 0$ which is then necessarily outside the $\rho(1)$st column; let $(2,\rho(2))$ be the leftmost position in the second row of A' with such an entry $\neq 0$ and call this entry u_2, then n-2 row operations of the type described in (9) will transform our matrix A' to a matrix A'' which has 0's only below as well as in front of u_1 ($\neq 0$ and in position $(1,\rho(1))$) and u_2 ($\neq 0$ and in position $(2,\rho(2))$). After repeating this procedure another (n-3)-times we will end up with a matrix

(16) $\qquad A^{(n-1)} = \prod_{j=n-1}^{1} \prod_{i=n}^{j+1} E_{ij}(-t_{ij})A$ such that if u_i is the leftmost entry $\neq 0$ in the i-th row of $A^{(n-1)}$, then u_i is in the $\rho(i)$-th column (where ρ is some permutation of n ciphers) and all entries below any u_i vanish.

Now set $T := \prod_{j=1}^{n-1} \prod_{i=j+1}^{n} E_{ij}(t_{ij}) = \begin{pmatrix} 1 & & 0 \\ & \ddots & \\ * & & 1 \end{pmatrix}$, $U := \begin{pmatrix} u_1 & & 0 \\ & \ddots & \\ 0 & & u_n \end{pmatrix}$ and $\pi := \rho^{-1}$

$\in S_n$ and conclude from (16) (using (10) and (11))

$$V := P(\pi)^{-1}U^{-1}T^{-1}A = P(\rho)U^{-1}A^{(n-1)} = \begin{pmatrix} 1 & & * \\ & \ddots & \\ 0 & & 1 \end{pmatrix}.$$

It remains to prove that $P(\pi)VP(\pi)^{-1}$ is upper triangular again; indeed, according to our construction, the matrix $U^{-1}A^{(n-1)}$ has the following properties (use (10) and (16)):

(17) the left-most entry $\neq 0$ in the i-th row equals 1 and is situated in the $\rho(i)$-th column; moreover, all entries below such an entry vanish.

Now our claim is obvious from (17) and (11) thanks to $P(\pi)VP(\pi)^{-1} = U^{-1}A^{(n-1)}P(\pi)^{-1}$ since transforming from $U^{-1}A^{(n-1)}$ to $U^{-1}A^{(n-1)}P(\pi)^{-1}$ moves the $\rho(i)$-th column to the i-th column. □

Example 1. In $M_2(D)$ the strict Bruhat normal form of a matrix looks as follows:

$$\begin{pmatrix} 0 & b \\ c & d \end{pmatrix} = \begin{pmatrix} 1 & 0 \\ db^{-1} & 1 \end{pmatrix}\begin{pmatrix} b & 0 \\ 0 & c \end{pmatrix}\begin{pmatrix} 0 & 1 \\ 1 & 0 \end{pmatrix}\begin{pmatrix} 1 & 0 \\ 0 & 1 \end{pmatrix} \text{ and}$$

$$\begin{pmatrix} a & b \\ c & d \end{pmatrix} = \begin{pmatrix} 1 & 0 \\ ca^{-1} & 1 \end{pmatrix}\begin{pmatrix} a & 0 \\ 0 & d-ca^{-1}b \end{pmatrix}\begin{pmatrix} 1 & 0 \\ 0 & 1 \end{pmatrix}\begin{pmatrix} 1 & a^{-1}b \\ 0 & 1 \end{pmatrix} \text{ if } a \neq 0.$$

Lemma 1. *Let* $V = \begin{pmatrix} 1 & & * \\ & \ddots & \\ 0 & & 1 \end{pmatrix}$ *and* $\pi \in S_n$ *be given; then we may find elements* $V' = \begin{pmatrix} 1 & & * \\ & \ddots & \\ 0 & & 1 \end{pmatrix}$ *and* $V'' = \begin{pmatrix} 1 & & * \\ & \ddots & \\ 0 & & 1 \end{pmatrix}$ *in* $M_n(D)$ *such that* $V = V'V''$ *and* $P(\pi)V''P(\pi)^{-1} = \begin{pmatrix} 1 & & \square \\ & \ddots & \\ 0 & & 1 \end{pmatrix}$ *as well as* $P(\pi)V'P(\pi)^{-1} = \begin{pmatrix} 1 & & 0 \\ & \ddots & \\ \square & & 1 \end{pmatrix}$.

Proof. Going through the existence proof of the previous theorem in the special case $A := P(\pi)V$ we see that we can write

$$P(\pi)V = A = T''P(\pi)V''$$

where the right hand side is the strict Bruhat normal form of the left hand side. Now set $V' := VV''^{-1}$; inspection shows that the required properties hold. □

Now we turn to a weakened version of Theorem 1; this new theorem turns out to be more suitable for applications (see next paragraph).

Theorem 2. *If a matrix* $A \in M_n(D)$ *(D a skew field) admits a decomposition*

(18) $A = TUP(\pi)V$ according to (12) but without the requirement that $P(\pi)VP(\pi)^{-1}$ be upper triangular, then the diagonal matrix U and the permutation π are still uniquely determined by A.

Proof. We start by writing $V = V'V''$ according to Lemma 1; this gives

(19) $A = TUP(\pi)V = TUT'P(\pi)V'' = T''UP(\pi)V''$

where

$T' := P(\pi)V'P(\pi)^{-1}$ and $T'' := TUT'U^{-1}$ are lower triangular (cf. Lemma 1 and (10)). Since the right-hand side of (19) is the strict Bruhat normal form of A the uniqueness of U and π is an immediate consequence of Theorem 1. □

Definition 2. Any decomposition of an $A \in M_n(D)$ according to (18) is called a "Bruhat normal form of A".

Theorem 3. Let D be a skew field and $A \in M_n(D)$, then the following conditions are equivalent:

(α) $A \in GL_n(D)$, i.e. A is invertible;
(β) $AB = 1$ for some $B \in M_n(D)$, i.e. A has a right inverse;
(γ) $CA = 1$ for some $C \in M_n(D)$, i.e. A has a left inverse;
(δ)   the rows of A are left linearly independent over D;
(ε)   the columns of A are right linearly independent over D.

Corollary 1. Precisely the matrices in $GL_n(D)$ (D a skew field) admit a Bruhat normal form and even the strict Bruhat normal form. □

Example 2. Let D be a ring (with $1 \neq 0$) which admits elements $a,b \in D$ such that $u := ab - ba \in D^*$ (e.g. take for D a skew field which is not commutative). Then

$$A := \begin{pmatrix} 1 & a \\ b & ab \end{pmatrix} = \begin{pmatrix} 1 & 0 \\ b & 1 \end{pmatrix}\begin{pmatrix} 1 & 0 \\ 0 & u \end{pmatrix}\begin{pmatrix} 1 & a \\ 0 & 1 \end{pmatrix} \in GL_n(D)$$

although the rows (resp. columns) of A are right (resp. left) linearly dependent over D. Consequently

$$A^t = \begin{pmatrix} 1 & b \\ a & ab \end{pmatrix} \notin GL_n(D) \quad \text{(cf. Exercise 1 in §2.)}. \text{ All this}$$

shows that *one must handle "right" and "left" in the context of* Theorem 3 *with the utmost care* !

Proof of Theorem 3. "(α)⇒(β),(γ)" is trivial; "(β),(δ)⇒(α)" follows from Theorem 1 because the various factors in the (strict) Bruhat normal form are invertible (use (1)..(6) in connection with (7)/(8) in §2.); "(γ)⇒(α)" is also clear, because the previous argument yields $C \in GL_n(D)$ and consequently $A \in GL_n(D)$; "$(\alpha) \Rightarrow (\delta), (\varepsilon)$" can be seen as follows: operations on a matrix according to (9),(10),(11) do not affect the properties described in items $(\delta)/(\varepsilon)$, but V, T visibly have these properties (if we decompose A according to Theorem 1); finally we must prove "$(\varepsilon) \Rightarrow (\alpha)$": indeed since $A \mapsto A^t$ defines an isomorphism $M_n(D)^{op} \simeq M_n(D^{op})$ (cf. (9) in §5.) we are able to make use of "$(\delta) \Rightarrow (\alpha)$" above. □

Exercise 1. Let D be a skew field with involution I (cf. §16.). Set $A := M_n(D)$ and extend I to A via $^I(a_{ij}) := (^I a_{ji})$. Now take $A \in A^*$ such that $^I A = A$ and let

$$A = TUP(\pi)V \text{ be a Bruhat normal form}, \quad U = \begin{pmatrix} u_1 & & 0 \\ & \ddots & \\ 0 & & u_n \end{pmatrix} ;$$

show
$$\pi^{-1} = \pi \text{ and } {}^I u_i = u_{\pi(i)} \quad (i = 1, \ldots, n)$$

(cf. P. Draxl [1980,p.108]).

Exercise 2. In the situation and with the notation of Exercise 1 show that one can achieve $V = {}^I T$ provided $char(D) \neq 2$.

Exercise 3. Use Exercise 2 for a proof of Theorem 5 in L. Elsner [1979].

§ 20. THE DIEUDONNÉ DETERMINANT

Thanks to the various results of the previous §19. the following definition is feasible.

Definition 1. *Let* D *be a skew field; define a function* $\delta\varepsilon\tau: M_n(D) \to D$ *by*

$$\delta\varepsilon\tau(A) := \begin{cases} 0 & \text{if} \quad A \notin GL_n(D), \\ \operatorname{sgn}(\pi)\prod_{i=1}^{n} u_i \neq 0 & \quad A \in GL_n(D); \end{cases}$$

here the u_i *are the (non vanishing) diagonal elements of the matrix* U *and* π *is the permutation of a Bruhat normal form of* A.

Note that U (hence the u_i) and π above are uniquely determined by A thanks to Theorem 2 in §19.

Example 1. Example 1 in §19. yields the following formulae in $M_2(D)$:

$$\delta\varepsilon\tau\begin{pmatrix} 0 & b \\ c & d \end{pmatrix} = -bc \quad \text{and} \quad \delta\varepsilon\tau\begin{pmatrix} a & b \\ c & d \end{pmatrix} = ad - aca^{-1}b \quad \text{if} \quad a \neq 0.$$

Let us establish rules for handling $\delta\varepsilon\tau$:

Lemma 1. *Let* D *be a skew field and* $B \in GL_n(D)$, *then*

$$\delta\varepsilon\tau\left(\begin{pmatrix} 1 & & 0 \\ & \ddots & \\ * & & 1 \end{pmatrix} B\right) = \delta\varepsilon\tau(B).$$

Proof. This is obvious since the matrix U and the permutation π in a Bruhat normal form of B remain unchanged. □

In order to avoid complicated and lengthy phrasing let us introduce the following (certainly not classical) notation:

Definition 2. *In a skew field* D *for elements* $a, b \in D$ *the notation* " $b = a\boxed{r}$ " *will stand for the phrase* " $a^{-1}b$ *is a product of at most* r *commutators of elements in* D^* ".

In this sense we claim

Lemma 2. *Let D be a skew field, $B \in GL_n(D)$ and $U := \begin{pmatrix} u_1 & & 0 \\ & \ddots & \\ 0 & & u_n \end{pmatrix}$
($u_i \neq 0$); then* $\delta\varepsilon\tau(UB) = \prod_{i=1}^{n} u_i \delta\varepsilon\tau(B)$ [n-1] .

Proof. Let $B = T'U'P(\pi')V'$ be a Bruhat normal form of B, then a straightforward calculation shows
$$UB = UT'U'P(\pi')V' = T''U''P(\pi')V'$$
where
$$T'' := UT'U^{-1} = \begin{pmatrix} 1 & & 0 \\ & \ddots & \\ * & & 1 \end{pmatrix} \text{ and } U'' = UU' = \begin{pmatrix} u_1'' & & 0 \\ & \ddots & \\ 0 & & u_n'' \end{pmatrix} \text{ (here } U' =$$
$$= \begin{pmatrix} u_1' & & 0 \\ & \ddots & \\ 0 & & u_n' \end{pmatrix} \text{ and } u_i'' = u_i u_i' \text{)}$$

This completes the proof thanks to $\prod_{i=1}^{n} u_i'' = \prod_{i=1}^{n} u_i \prod_{i=1}^{n} u_i'$ [n-1] . □

Now we come to the core of our analysis of the function $\delta\varepsilon\tau$.

Lemma 3. *Let D be a skew field, $B \in GL_n(D)$ and $P(i,i+1)$ the permutation matrix belonging to the transposition $(i\ i+1)$; then*
$$\delta\varepsilon\tau(P(i,i+1)B) = -\delta\varepsilon\tau(B) \boxed{1} .$$

Proof. Let $B = TUP(\pi)V$ be a Bruhat normal form of B and write $T = (t_{rs})$ (with $t_{rr} = 1$ and $t_{rs} = 0$ if $r < s$); now denote $t :=$
$:= u_{i+1}^{-1} t_{i+1,i} u_i \in D$ where u_r is the r-th entry of the (diagonal) matrix U in the above Bruhat normal form. A straightforward calculation gives

(1) $P(i,i+1)B = \begin{pmatrix} 1 & & 0 \\ & \ddots & \\ * & & 1 \end{pmatrix} \begin{pmatrix} \ddots & & 0 \\ & u_{i+1} & \\ & & u_i \\ 0 & & \ddots \end{pmatrix} E_{i,i+1}(t) P(i,i+1) P(\pi) V$.

Now assume *either* $t = 0$ (this amounts to $t_{i+1,i} = 0$) *or* $\pi^{-1}(i) > \pi^{-1}(i+1)$ (this amounts to the fact that $E_{i,i+1}(t)$ remains upper triangular after being "moved around the two permutation matrices" on its right hand side in (1)); then it follows (cf. (1)..(11) in §19.) from (1):

(2) $P(i,i+1)B = \begin{pmatrix} 1 & & 0 \\ & \ddots & \\ * & & 1 \end{pmatrix} \begin{pmatrix} \ddots & & 0 \\ & u_{i+1} & \\ & & u_i \\ 0 & & \ddots \end{pmatrix} P((i\ i+1)\pi) \begin{pmatrix} 1 & & * \\ & \ddots & \\ 0 & & 1 \end{pmatrix}$.

Since the right-hand side of (2) is a Bruhat normal form of the left-hand side thereof and since $sgn(i\ i+1) = -1$, we get our assertion immediately from $u_1 \cdots u_{i+1} u_i \cdots u_n = \prod_{r=1}^{n} u_r$ [1] .

Now assume the contrary of the above, i.e. $t \neq 0$ (this amounts to $0 \neq t_{i+1,i}$) and $\pi^{-1}(i) < \pi^{-1}(i+1)$ (this amounts to the fact that the matrix $E_{i,i+1}(t)$ becomes lower trianguler after being "moved around..." (see above) in (1)); an easy inspection shows that the identity

$$\begin{pmatrix} 1 & t \\ 0 & 1 \end{pmatrix}\begin{pmatrix} 0 & 1 \\ 1 & 0 \end{pmatrix} = \begin{pmatrix} 1 & 0 \\ t^{-1} & 1 \end{pmatrix}\begin{pmatrix} t & 0 \\ 0 & -t^{-1} \end{pmatrix}\begin{pmatrix} 1 & t^{-1} \\ 0 & 1 \end{pmatrix}$$

is valid, and this gives the formula ($t \in D^*$)

(3) $\quad E_{i,i+1}(t)P(i,i+1) = E_{i+1,i}(t^{-1})D_i(t)D_{i+1}(-t^{-1})E_{i,i+1}(t^{-1})$

(note that the right-hand side of (3) is the strict Bruhat normal form of its left-hand side). Merging (1) with (3) yields (after using (1)..(11) in §19. several times)

(4) $\quad P(i,i+1)B = \begin{pmatrix} 1 & & 0 \\ & \ddots & \\ * & & 1 \end{pmatrix}\begin{pmatrix} \ddots & & 0 \\ & u_{i+1}t & \\ & -u_i t^{-1} & \\ 0 & & \ddots \end{pmatrix} P(\pi) \begin{pmatrix} 1 & & \square \\ & \ddots & \\ 0 & & 1 \end{pmatrix}$

where the right-hand side of (4) is a Bruhat normal form of the left-hand side thereof. This gives the required formula for $\delta\varepsilon\tau$ thanks to

$u_1..(-u_{i+1}tu_i t^{-1})..u_n = -u_1..t_{i+1,i}u_i t_{i+1,i}^{-1}u_{i+1}..u_n = -\prod_{r=1}^{n} u_r \boxed{1}$. □

The three preceding lemmas may be united into a single statement, namely

Theorem 1 . *Let* D *be a skew field and* $\delta\varepsilon\tau: M_n(D) \to D$ *the function according to* Definition 1; *then* $\delta\varepsilon\tau$ *is surjective and has the property*

$$\delta\varepsilon\tau(AB) = \begin{cases} \delta\varepsilon\tau(A)\delta\varepsilon\tau(B) & \text{if} \quad A \notin GL_n(D) \text{ or } B \notin GL_n(D), \\ \delta\varepsilon\tau(A)\delta\varepsilon\tau(B)\boxed{n^2-1} & A,B \in GL_n(D) . \end{cases}$$

Corollary 1 . *Let* D *be a skew field and* $\delta\varepsilon\tau$ *as above; then* $\delta\varepsilon\tau$ *induces a surjective homomorphism of groups*

$$\det: GL_n(D) \longrightarrow D^*/[D^*,D^*] =: D^{*ab}$$

which coincides with the usual determinant in case D *is a commutative field.*

Definition 3 . *The function* det *according to* Corollary 1 *is called the "Dieudonné determinant".*

Proof. The upper line of the formula to be proved is clear, also the surjectivity of $\delta\varepsilon\tau$ thanks to $\delta\varepsilon\tau(D_i(d)) = d$ for any index i ($d \in D$). For the proof of the lower line of our formula let $A = TUP(\pi)V$ be a Bruhat normal form of A , hence

$$\delta\varepsilon\tau(A) = \text{sgn}(\pi)\prod_{i=1}^{n} u_i \quad \text{where} \quad U = \begin{pmatrix} u_1 & & 0 \\ & \ddots & \\ 0 & & u_n \end{pmatrix} .$$

Now select the permutation $\tau \in S_n$ such that $P(\tau) = P(\pi)\begin{pmatrix} 0 & 1 \\ 1 & 0 \end{pmatrix}$, hence

$\text{sgn}(\tau) = (-1)^{(n-1)n/2}\text{sgn}(\pi)$ since the matrix $\begin{pmatrix} 0 & 1 \\ 1 & 0 \end{pmatrix}$ is a product of $(n-1)n/2$ matrices of the type $P(i,i+1)$. Therefore (cf. (7) in §19.)

(5)
$$A = TUP(\tau)T'\begin{pmatrix} 0 & 1 \\ 1 & 0 \end{pmatrix} \quad \text{where } T' \text{ arises from } V \text{ via rotation by 180 degrees (and is hence lower triangular with 1's in the main diagonal),}$$

and consequently (using Lemmas 1/2/3 repeatedly in connection with (5))

$$\delta\varepsilon\tau(AB) = \delta\varepsilon\tau(TUP(\tau)T'\begin{pmatrix} 0 & 1 \\ 1 & 0 \end{pmatrix}B) = \delta\varepsilon\tau(UP(\tau)T'\begin{pmatrix} 0 & 1 \\ 1 & 0 \end{pmatrix}B) =$$

$$= \prod_{i=1}^{n} u_i \delta\varepsilon\tau(P(\tau)T'\begin{pmatrix} 0 & 1 \\ 1 & 0 \end{pmatrix}B) \boxed{n-1} =$$

$$= \text{sgn}(\tau)\prod_{i=1}^{n} u_i \delta\varepsilon\tau(T'\begin{pmatrix} 0 & 1 \\ 1 & 0 \end{pmatrix}B) \boxed{(n-1)+(n-1)n/2} =$$

$$= \text{sgn}(\tau)(-1)^{(n-1)n/2}\prod_{i=1}^{n} u_i \delta\varepsilon\tau(B) \boxed{(n-1)(1+n)} =$$

$$= \text{sgn}(\pi)\prod_{i=1}^{n} u_i \delta\varepsilon\tau(B) \boxed{(n-1)(n+1)} = \delta\varepsilon\tau(A)\delta\varepsilon\tau(B) \boxed{n^2-1} \quad . \square$$

Corollary 2. *If D is a skew field and $A, B \in GL_n(D)$, then*

$$\delta\varepsilon\tau(B^{-1}AB) = \delta\varepsilon\tau(B)^{-1}\delta\varepsilon\tau(A)\delta\varepsilon\tau(B) \boxed{2(n^2-1)} \quad .$$

Proof. Apply Theorem 1 to both sides of the equation $B(B^{-1}AB) = AB$. □

Corollary 3. *If D is a skew field and $A \in GL_n(D)$ a product of at most r commutators of elements in $GL_n(D)$, then*

$$\delta\varepsilon\tau(A) = \boxed{r+(4r-1)(n^2-1)} \quad \text{and}$$
$$\delta\varepsilon\tau([GL_n(D), GL_n(D)]) = [D^*, D^*] \quad .$$

Proof. The second statement follows from the first one; as for the first one, apply Corollary 2 to both sides of the equation $A(A^{-1}B^{-1}AB) = B^{-1}AB$ which yields our assertion in case $r = 1$. Then use induction on r. □

Definition 4. *Let D be a skew field; write $SL_n(D) := \text{Ker det} \trianglelefteq GL_n(D)$ and $E_n(D) := \langle E_{ij}(t) \mid 1 \leq i,j \leq n, i \neq j, t \in D \rangle \leq GL_n(D)$*

Obviously $E_n(D) \subseteq SL_n(D)$; our goal is to prove equality. For this purpose we need two auxiliary results which are also of general interest.

Theorem 2. *Let D be a skew field and $A \in GL_n(D)$, then $A = ED_n(a)$ for suitable $E \in E_n(D)$ and $a \in D^*$ where the class $\bar{a} = \det(A) \in D^{*ab}$ is uniquely determined by A.*

Corollary 4. *Let D be a skew field, then every matrix $A \in GL_n(D)$ is a product of triangular matrices.*

Proof. The corollary and the last assertion of the theorem are clear; it remains to prove the existence of such a decomposition $A = ED_n(a)$. We proceed by repeated row operations of type (9) in §19. as follows: thanks to Theorem 3 in §19. the first column of A must have an entry $\neq 0$, hence we can transform A via left multiplication with elements from $E_n(D)$ to a matrix $A' \in GL_n(D)$ such that A' has a 1 in position (1,1) and 0's elsewhere in the first column (note that we do *not* need row operations of type (10) in §19.; cf. also E. Artin [1957,pp.151]).
Now we observe (again because of Theorem 3 in §19.) that the second column of A' must have an entry $\neq 0$ outside the first row, hence we can transform A' via left multiplication with elements from $E_n(D)$ to a matrix $A'' \in GL_n(D)$ with 1's in positions (1,1) and (2,2) and with 0's elsewhere in the first two columns. After repeating this procedure another $n-2$ times we end up with a matrix $D_n(a)$ for some $a \in D^*$.
Summarizing we find

$\qquad FA = D_n(a)$ with some matrix F which is a product of certainly less than n^2 matrices of type $E_{ij}(t)$,

hence

$\qquad A = ED_n(a)$ where $E := F^{-1} \in E_n(D)$. $\square$

Whitehead Lemma. *Let* R *be a ring (with* $1 \neq 0$ *) and* $u,v \in R^*$, *then*

$$\begin{pmatrix} u & 0 \\ 0 & u^{-1} \end{pmatrix}, \begin{pmatrix} 1 & 0 \\ 0 & uvu^{-1}v^{-1} \end{pmatrix} \in E_2(R) \ .$$

Note that the definition of $E_n(R)$ (cf. Definition 4) makes also sense for rings (and not only for skew fields).

Proof. Inspection shows that we have the identities

$$\begin{pmatrix} u & 0 \\ 0 & u^{-1} \end{pmatrix} = \begin{pmatrix} 1 & 1 \\ 0 & 1 \end{pmatrix}\begin{pmatrix} 1 & 0 \\ -1 & 1 \end{pmatrix}\begin{pmatrix} 1 & 1 \\ 0 & 1 \end{pmatrix}\begin{pmatrix} 1 & -u^{-1} \\ 0 & 1 \end{pmatrix}\begin{pmatrix} 1 & 0 \\ u & 1 \end{pmatrix}\begin{pmatrix} 1 & -u^{-1} \\ 0 & 1 \end{pmatrix}$$

and

$$\begin{pmatrix} 1 & 0 \\ 0 & uvu^{-1}v^{-1} \end{pmatrix} = \begin{pmatrix} u^{-1} & 0 \\ 0 & u \end{pmatrix}\begin{pmatrix} v^{-1} & 0 \\ 0 & v \end{pmatrix}\begin{pmatrix} vu & 0 \\ 0 & u^{-1}v^{-1} \end{pmatrix} \ . \quad \square$$

Theorem 3. *If* D *is a skew field, then* $E_n(D) = SL_n(D)$.

Proof. "$\subseteq$" has been observed already; as for "$\supseteq$" everything is clear from Theorem 2 because the Whitehead Lemma obviously implies $D_n(a) \in E_n(D)$ for all $a \in [D^*,D^*]$ $\square$

Definition 5. *Let* A *be a simple ring* $\neq M_2(F_2)$, *then*
$\qquad K_1(A) := A^*/[A^*,A^*] =: A^{*ab}$
is called the "Whitehead group of A ".

It should be pointed out that in general $K_1(R)$ may be defined *for any ring* R (with $1 \neq 0$) ; for this see e.g. J. Milnor [1971]. In the case of a simple ring $\neq M_2(F_2)$ our definition is easier to state and amounts to the same.

Theorem 4. *Let* D *be a skew field, then*
(i) $SL_n(D) = [GL_n(D), GL_n(D)]$ *unless* $n = 2$ *and* $D = F_2$;
(ii) $SL_n(D) = [SL_n(D), SL_n(D)]$ *unless* $n = 2$ *and* $D = F_2$ *or* F_3 ;
(iii) *if* $A := M_n(D) \neq M_2(F_2)$, *then the Dieudonné determinant* det *induces an isomorphism*
$$K_1(\det): K_1(A) \xrightarrow{\sim} K_1(D) ;$$
its inverse is induced by $a \mapsto D_n(a)$.

Proof. *(iii)* is clear from *(i)* together with Corollary 1 and Theorems 2/3. Moreover, "$\supseteq$" in *(i)* and *(ii)* is obvious; on the other hand we have in $GL_2(F_3)$ the identity

$$\begin{pmatrix} 1 & 1 \\ 0 & 1 \end{pmatrix} = \begin{pmatrix} 1 & -2 \\ 0 & 1 \end{pmatrix} = \begin{pmatrix} 1 & 0 \\ 0 & -1 \end{pmatrix} \begin{pmatrix} 1 & 1 \\ 0 & 1 \end{pmatrix} \begin{pmatrix} 1 & 0 \\ 0 & -1 \end{pmatrix}^{-1} \begin{pmatrix} 1 & 1 \\ 0 & 1 \end{pmatrix}^{-1} ,$$

hence — after eventually going over to transpose and inverse —
$$SL_2(F_3) \subseteq [GL_2(F_3), GL_2(F_3)] .$$
Therefore it suffices to show "$\subseteq$" in *(ii)*; by Theorem 3 this is clear thanks to (3) in §19. provided $n \geq 3$, i.e. we are left with the case $n = 2$ and $|D| \geq 4$. Here we make use of the identities

$$\begin{pmatrix} 1 & b-aba \\ 0 & 1 \end{pmatrix} = \begin{pmatrix} a & 0 \\ 0 & a^{-1} \end{pmatrix} \begin{pmatrix} 1 & -b \\ 0 & 1 \end{pmatrix} \begin{pmatrix} a & 0 \\ 0 & a^{-1} \end{pmatrix}^{-1} \begin{pmatrix} 1 & -b \\ 0 & 1 \end{pmatrix}^{-1}$$

and

$$\begin{pmatrix} 1 & 0 \\ b-aba & 1 \end{pmatrix} = \begin{pmatrix} a^{-1} & 0 \\ 0 & a \end{pmatrix} \begin{pmatrix} 1 & 0 \\ -b & 1 \end{pmatrix} \begin{pmatrix} a^{-1} & 0 \\ 0 & a \end{pmatrix}^{-1} \begin{pmatrix} 1 & 0 \\ -b & 1 \end{pmatrix}^{-1}$$

which show — in connection with the Whitehead Lemma — that it suffices to prove the following technical result:

Lemma 4. *Let* D *be a skew field* $\neq F_2, F_3$; *then every* $d \in D$ *may be written*
$$d = \sum_{r=1}^{4} (a_r b_r a_r^{-1} b_r) \text{ with suitable } a_r, b_r \in D^* .$$

Proof. In case "$\text{char}(D) \neq 2$" (this implies $|D| \geq 5$ and $\frac{1}{2} \in Z(D)$) we can write *either*
$$d = \frac{d+1}{2} 1 \frac{d+1}{2} - 1 + \frac{d-1}{2}(-1)\frac{d-1}{2} - (-1) \text{ if } d \neq 1, -1$$
or
$$\mp 1 = \mp(d+1) + (\pm d) \text{ with some } d \neq -2, -1, 0, 1 \text{ where } \mp(d+1), \pm d$$
may be written as in the first of these two subcases.

In the remaining case "$\text{char}(D) = 2$" (i.e. $1 = -1$) we write *either*
$$1 = (d+1)1(d+1) - 1 + d1d - 1 \text{ with some } d \neq 0, 1 \text{ or}$$

$$0 = 1^3 - 1 \quad \textit{or finally}$$
$$d = (d+1)d(d+1) - d + d^3 - d \quad \text{if} \quad d \neq 0,1 \, . \, \square \, \square$$

We close this paragraph with a few additional remarks concerning Theorem 4: of course one would wish to know how many commutators are really necessary in *(i)* and *(ii)* in order to represent an element in $SL_n(D)$. At least in the case of a commutative field D this question has a nice answer: "one" in *(i)* and "at most two" in *(ii)* (cf. R.C. Thompson [1961]). In the non commutative case, however, all this turns out to be much more complicated; references are e.g. U. Rehmann [1980] and В.В. Курсов [1979] in the (generally not so well-known) reports of the Byelorussian Academy of Sciences at Minsk (USSR).

Exercise 1. In the situation and with the notation of Exercise 1 in §19. show: if $A \in \mathbb{A}$ is such that $^I A = A$, then $\delta \varepsilon \tau(A)$ is modulo $[D^*, D^*]$ a product of at most n elements $d \in D$ satisfying $^I d = d$.

§ 21. THE STRUCTURE OF $SL_n(D)$ FOR $n \geq 2$

In the previous paragraph (cf. Theorem 4 *ibid.*) we have seen that $SL_n(D)$ is a *perfect group* (i.e. coincides with its own commutator subgroup) provided D is a skew field (unless $n = 2$ *and* $D = F_2$ or $= F_3$). Here we aim at a deeper investigation of the structure of $SL_n(D)$. For an alternative point of view the reader should consult Chapter IV in E. Artin [1957] as well as Chapter II *ibid.* where the relationship of all that to projective and affine geometry is emphasized much more than here.

Theorem 1. *Let D be a skew field, then*

$$Z(GL_n(D)) = \{ d1 \mid d \in Z(D)^* \} \text{ and }$$

$$Z(SL_n(D)) = \{ d1 \mid d \in Z(D)^* \text{ and } d^n \in [D^*,D^*] \}.$$

Proof. According to Lemma 2 in §2. the right-hand sides in the theorem are equal to $Z(M_n(D)) \cap GL_n(D)$ resp. $Z(M_n(D)) \cap SL_n(D)$, hence we get "$\supseteq$" of our claim. For the proof of the converse it clearly suffices to show

$$\{ A \in M_n(D) \mid AB = BA \text{ for all } B \in SL_n(D) \} \subseteq Z(M_n(D)) .$$

By virtue of Theorem 3 in §20. this is exactly what has been shown in Lemma 2 in §2. ◻

In the sequel we shall study three auxiliary results.

Lemma 1. *Let D be a skew field and* $G := SL_n(D)$ *(resp.* $:= GL_n(D)$ *), then* $G_i := \{ (a_{rs}) \in G \mid a_{is} = 0 \text{ for all } s \neq i \}$ *is a maximal subgroup in G* ($i = 1,..,n$; $n \geq 2$).

Proof. The fact that G_i is a group is checked quickly by inspection. Now by definition (cf. §§2./19.)

(1) $\qquad E_{rs}(t) \in G_i$ (resp. $E_{rs}(t), D_s(u) \in G_i$) for all $r \neq i$,

hence by virtue of Theorem 3 in §20. (resp. Theorems 2/3 in §20.) using (3) in §19. it suffices to show the following:

(2) given a subgroup $H \subseteq G$ such that $H \neq G_i$, $H \supset G_i$; then $E_{is}(t) \in H$ for some $s \neq i$ and all $t \in D^*$, hence $H = G$.

For the proof of (2) let us select a matrix $A = (a_{rs}) \in H \setminus G_i$; this implies $a_{is} \neq 0$ for some index $s \neq i$. After possible left multiplication with $D_i(ta_{is}^{-1})D_s(a_{is}t^{-1}) \in SL_n(D)$ we may assume that we can find an index $s \neq i$ with the following property:

given $t \in D^*$, then there exists a matrix $A = (a_{rs}) \in$
$\in H \setminus G_i$ such that $a_{is} = t$.

From now on we proceed basically as in the proof of Theorem 2 in §20., namely: by repeated use of operations of type (9) in §19. we may transform from A to $E_{is}(t)$ inside our subgroup H, hence (2). □

Lemma 2. *If in the situation of Lemma 1 we have a normal subgroup $N \triangleleft G$ such that $N \not\subseteq Z(G)$, then $NG_i = G$ for some index i.*

Proof. If $N \not\subseteq G_i$ for some index i, then $NG_i = G$ by the preceding lemma. Now suppose $N \subseteq \bigcap_{i=1}^{n} G_i = \{\text{diagonal matrices in } G\}$; since N is normal in G the matrix (cf. (9)/(10) in §19.)

$$E_{rs}(t)\begin{pmatrix} u_1 & & 0 \\ & \ddots & \\ 0 & & u_n \end{pmatrix}E_{rs}(t)^{-1} = \begin{pmatrix} u_1 & & 0 \\ & \ddots & \\ 0 & & u_n \end{pmatrix}E_{rs}(u_r^{-1}tu_s - t)$$

must be a diagonal matrix in G for all $r \neq s$ and all $t \in D$, hence in particular $0 \neq u_1 = .. = u_n \in Z(D)$, i.e. $N \subseteq Z(G)$ by Theorem 1. □

Lemma 3. *Let D be a skew field, $G := SL_n(D)$ and G_i according to Lemma 1; then $H_i := \langle E_{ri}(t) \mid t \in D ; r = 1,..,n ; r \neq i \rangle$ is a normal abelian subgroup of G_i such that ($i = 1,..,n$; $n \geq 2$)*

(3) $\quad G = \langle gH_ig^{-1} \mid g \in G \rangle$ *(i.e. G is the normal closure of any of its subgroups H_i).*

Proof. The fact that H_i is an abelian group is clear from (2) in §19. Thanks to (3) in §19. the assertion (3) is certainly true in case $n \geq 3$ (cf. also Theorem 3 in §20.). In the remaining case $n = 2$, however, (3) is also easily seen because of

$$E_{12}(t) = \begin{pmatrix} 0 & -1 \\ 1 & 0 \end{pmatrix}^{-1} E_{21}(-t) \begin{pmatrix} 0 & -1 \\ 1 & 0 \end{pmatrix}.$$

Therefore we are left with showing that H_i is normal in G_i: let $A = (a_{rs}) \in G_i$ and $B := A^{-1} = (b_{mr}) \in G_i$, then (cf. (9) in §19.) $E_{ji}(t)A =: A' = (a'_{rs})$ where $a'_{ji} = a_{ji} + ta_{ii}$ and $a'_{rs} = a_{rs}$ in all other cases. It follows

$$\sum_{r=1}^{n} b_{mr} a'_{rs} = \begin{cases} = 1 \text{ if } m = s \neq i \text{ and } = 0 \text{ if } m \neq s \neq i, \\ = 1 \text{ if } m = s = i \text{ and } = b_{mj} ta_{ii} \text{ if } m \neq s = i \end{cases}$$

which amounts to

$$A^{-1} E_{ji}(t) A = BA' = \prod_{\substack{m=1 \\ m \neq i}}^{n} E_{mi}(b_{mj} ta_{ii}) \in H_i \quad (j \neq i). \quad \square$$

Now we come to the principal result of this paragraph.

Theorem 2. *Let D be a skew field, then – unless $n = 2$ and $D = F_2$ or $= F_3$ – the proper normal subgroups of $SL_n(D)$ ($n \geq 2$) are precisely the subgroups of the centre $Z(SL_n(D))$, i.e. the factor group*

$$PSL_n(D) := SL_n(D)/Z(SL_n(D)) \text{ is a simple group for } n \geq 2.$$

Proof. Let $N \trianglelefteq SL_n(D) =: G$ be a normal subgroup such that $N \not\subseteq Z(G)$; then Lemma 2 implies $NG_i = G$ for some index i, hence (because of Lemma 3) we obtain

$$G = \langle n g_i H_i g_i^{-1} n^{-1} \mid n \in N, g_i \in G_i \rangle = \langle n H_i n^{-1} \mid n \in N \rangle = NH_i$$

and therefore – since H_i is abelian – $[G,G] = [NH_i, NH_i] \subseteq N$. Now use *(ii)* of Theorem 4 in §20. which implies $G = [G,G] = N$. $\square$

Theorem 2 leads to a *series of finite simple groups* $PSL_n(F_q)$ (unless $n = 2$ and $q = 2$ or $= 3$); remarks on these may be found e.g. in E. Artin [1957, pp.170]. Another good reference for this paragraph is Chapter II in J. Dieudonné [1963].

Exercise 1. Show that $GL_2(F_2) = SL_2(F_2) = PSL_2(F_2)$ is isomorphic to the *dihedral group* of order 6 (which is neither perfect nor simple).

Exercise 2. Prove that $SL_2(F_3)$ is not perfect and show that $PSL_2(F_3)$ is isomorphic to the *alternating group* of degree 4 (which has order 12 and is not simple).

Note that the exercises show that Theorem 2 is best possible.

§ 22 . REDUCED NORMS AND TRACES

Let A be an (associative) K-algebra (K a (commutative) field) of finite degree $|A:K| =: m$. Consider the left multiplication
$$L_a : A \to A \, , \, x \mapsto ax$$
and identify it (as usual) via $L_a \in \text{End}_K(A) \simeq M_m(K)$ with a matrix.

Definition 1 . *In the above situation*

$$N_{A/K}(a) := \det(L_a) \qquad \text{\textit{is called the}} \quad \begin{array}{l}\text{"norm of } A/K \text{ "}\\ \text{"trace of } A/K \text{ "}\end{array} \qquad (a \in A) .$$
$$Tr_{A/K}(a) := tr(L_a)$$

Of course, the above norm resp. trace coincide with the usual norm resp. trace in case A/K is a (finite) field extension. Moreover, the following formulae are clear from the definitions (since the corresponding properties of det and tr are well-known from Linear Algebra):

(1) $\quad N_{A/K} : A \to K$ is a multiplicative map ;
(2) $\quad N_{A/K}(a) = a^m$ if $a \in K$ ($m = |A:K|$) ;
(3) $\quad Tr_{A/K} : A \to K$ is K-linear ;
(4) $\quad Tr_{A/K}(a) = ma$ if $a \in K$ ($m = |A:K|$) ;
(5) $\quad Tr_{A/K}(aa') = Tr_{A/K}(a'a)$ ($a,a' \in A$) .

Furthermore, we find

(6) $\quad N_{A/K}(a) \neq 0$ if and only if $a \in A^*$;

here the "if" is obvious whereas the "only if" may be seen as follows: $N_{A/K}(a) \neq 0$ means that L_a is an automorphism, hence $L_a f = \text{id}_A$ for some $f \in \text{Aut}_K(A)$, i.e. $af(1) = 1$ and therefore $a \in A^*$.

The next result should be well-known:

Lemma 1 . *Let L/K be a finite field extension of degree $m := |L:K|$ and V a vector space over L of finite dimension $n := \dim_L(V)$ (hence $mn = \dim_K(V)$). Then any $f \in \text{End}_L(V)$ may be viewed as an element in $\text{End}_K(V)$*

and in this sense we have

$$\det_K(f) = N_{L/K}(\det_L(f)) \quad and \quad tr_K(f) = Tr_{L/K}(tr_L(f)) \ .$$

Proof. Let $\{e_1,..,e_n\}$ be an L-basis of V and $\{f_1,..,f_m\}$ a K-basis of L, hence $\{f_i e_j \mid 1 \leq i \leq m, 1 \leq j \leq n\}$ a K-basis of V. Then we get

$$f(e_j) = \sum_{r=1}^{n} x_{rj} e_r \quad \text{with} \quad x_{rj} \in L \quad (1 \leq j \leq n) \quad and$$

$$L_{x_{rj}}(f_i) = \sum_{s=1}^{m} y_{si}^{rj} f_s \quad \text{with} \quad y_{si}^{rj} \in K \quad (1 \leq i \leq m) \ ,$$

hence

$$f(f_i e_j) = f_i f(e_j) = \sum_{r=1}^{n} \sum_{s=1}^{m} y_{si}^{rj} f_s e_r \ .$$

All this implies the second of our two formulae thanks to

$$tr_K(f) = \sum_{r=1}^{n} \sum_{s=1}^{m} y_{ss}^{rr} = \sum_{r=1}^{n} Tr_{L/K}(x_{rr}) = Tr_{L/K}(\sum_{r=1}^{n} x_{rr}) =$$
$$= Tr_{L/K}(tr_L(f)) \ .$$

The first formula is somewhat deeper; for its proof we apply the following trick: on both sides of the formula we find multiplicative functions, hence - by Corollary 4 in §20. - it suffices to assume that

$$(x_{rj}) \in GL_n(L) \quad \text{is triangular}$$

(note that the case $f \notin \text{Aut}_L(V)$ is trivial since both sides of our formula vanish). Then

$$(y_{si}^{rj}) \in M_n(M_m(K)) = M_{mn}(K) \quad \text{is block triangular}$$

with the (m,m)-blocks $(y_{si}^{rr}) \in M_m(K)$ in the main block diagonal. Now we may use standard results from Linear Algebra concerning determinants of triangular block matrices and obtain

$$\det_K(f) = \prod_{r=1}^{n} \det_K(y_{si}^{rr}) = \prod_{r=1}^{n} N_{L/K}(x_{rr}) = N_{L/K}(\prod_{r=1}^{n} x_{rr}) =$$
$$= N_{L/K}(\det_L(f)) \quad \square$$

An immediate application of Lemma 1 are the

Tower Formulae . *Let* L/K *be a finite field extension and* A *a finite dimensional L-algebra, then* ($a \in A$)

$$N_{A/K}(a) = N_{L/K}(N_{A/L}(a)) \quad and \quad Tr_{A/K}(a) = Tr_{L/K}(Tr_{A/L}(a)) \ . \quad \square$$

In the case of a central simple K-algebra A it turns out that $N_{A/K}$ and $Tr_{A/K}$ are not sophisticated enough (cf. Lemma 3 below); in order to understand this we need some preparation: in what follows we use the notation and terminology from Part II and start our investigations with the highly important

Lemma 2. *Let* $[A] \in Br(K)$ *and* E *a splitting field of* A, *i.e. one has an isomorphism* $f: A \otimes_K E \xrightarrow{\sim} M_n(E)$. *Then we have*

$$\det(f(a \otimes 1)), tr(f(a \otimes 1)) \in K \text{ for all } a \in A$$

and these elements in K *are independent of the choice of* E *and* f.

Definition 2. *In the above situation* ($a \in A$)

$$RN_{A/K}(a) := \det(f(a \otimes 1)) \qquad \text{"reduced norm of } A/K\text{"}$$
$$\text{is called the}$$
$$RTr_{A/K}(a) := tr(f(a \otimes 1)) \qquad \text{"reduced trace of } A/K\text{"}.$$

Proof. If we have another isomorphism $g: A \otimes_K E \xrightarrow{\sim} M_n(E)$ then $g(a \otimes 1) =$
$= tf(a \otimes 1)t^{-1}$ for some $t \in GL_n(E)$ because of the Skolem-Noether Theorem in §7., hence our claim. Now let F be another splitting field of A; we may assume $E \subseteq F$ (for otherwise consider EF and apply our reasoning twice in the situations $E \subseteq EF$ and $F \subseteq EF$) and consider the isomorphisms

$$A \otimes_K F \xrightarrow[h]{\sim} (A \otimes_K E) \otimes_E F \xrightarrow[f \otimes id_F]{\sim} M_n(E) \otimes_E F \xrightarrow[j_F]{\sim} M_n(F)$$

where h is according to Lemma 2 in §5. (such that $h(a \otimes x) = (a \otimes 1) \otimes x$ for all $a \in A$, $x \in F$) and j_F is the isomorphism of Lemma 4 in §5. Now call $g := j_F(f \otimes id_F)h$; it follows

$$g(a \otimes 1) = j_F(f(a \otimes 1) \otimes 1) = f(a \otimes 1) \quad \text{for all} \quad a \in A,$$

hence the claimed independence of the choice of the splitting field. It remains to be proved that $\det(f(a \otimes 1))$ and $tr(f(a \otimes 1))$ lie in the base field K. Now by what we have just seen we may assume E/K to be finite Galois (cf. Theorem 9 in §9.) with $\Gamma := Gal(E/K)$; then $A \otimes_K E$ resp. $M_n(E)$ carry the structure of a left Γ-module via

$$^\sigma x := id_A \otimes \sigma(x) \quad (x \in A \otimes_K E) \text{ resp. componentwise in } M_n(E).$$

Therefore it makes sense to define a map $^\sigma f$ by

$$^\sigma f: A \otimes_K E \to M_n(E), \quad x \mapsto {}^\sigma(f({}^{\sigma^{-1}} x)) \quad (\sigma \in \Gamma).$$

A straightforward calculation shows that $^\sigma f$ is likewise an E-algebra isomorphism for every $\sigma \in \Gamma$, hence - since the independence of the choice of the isomorphism is already known -

$$\det(f(a \otimes 1)) = \det({}^\sigma f(a \otimes 1)) = \det({}^\sigma(f(a \otimes^{\sigma^{-1}} 1))) =$$
$$= \det({}^\sigma(f(a \otimes 1))) = {}^\sigma \det(f(a \otimes 1)) \quad \text{for all} \quad \sigma \in \Gamma$$

which amounts to $\det(f(a \otimes 1)) \in K$ ($tr(f(a \otimes 1)) \in K$ is shown similarly). □

Lemma 3. *Let* $[A] \in Br(K)$ *and* $n^2 := |A:K|$ *(i.e.* n *the reduced degree of* A/K *), then*

$$N_{A/K}(a) = (RN_{A/K}(a))^n \quad and \quad Tr_{A/K}(a) = n(RTr_{A/K}(a)) \quad (a \in A).$$

This lemma, incidentally, justifies the names *reduced* norm resp. trace !

Proof. Select a splitting field E of A such that $|E:K|$ divides n (cf. Corollary 6 in §9.), i.e. $rs = n$ where $r := |E:K|$. Now consider the maps

$$\begin{array}{ccccccccc} a \otimes 1 & A \otimes_K E & \xrightarrow{f} & M_n(E) & \simeq \text{End}_E(E^n) & \subseteq \text{End}_K(E^n) & \simeq & M_{rn}(K) & A \\ \uparrow & \uparrow & & & & & & \downarrow & \downarrow \\ a & A & - & - & - & \xrightarrow{h} & & M_{n^2}(K) = M_s(M_{rn}(K)) & A1 \end{array}$$

Thanks to Lemmas 1/2 we find

$$\det(h(a)) = (\det_K(f(a \otimes 1)))^s = (N_{L/K}(\det_L(f(a \otimes 1))))^s =$$
$$= (\det_L(f(a \otimes 1)))^n = (RN_{A/K}(a))^n \quad .$$

On the other hand (by Definition 1) we get $N_{A/K}(a) = \det(h(a))$ thanks to the Skolem-Noether Theorem in §7., hence our first claim; the second claim (concerning traces) follows in exactly the same manner. □

Now we state five formulae which are more or less clear from our definitions:

(7) $RN_{A/K}: A \to K$ is a multiplicative map ;
(8) $RN_{A/K}(a) = a^n$ if $a \in K$ ($n^2 = |A:K|$)
(9) $RTr_{A/K}: A \to K$ is K-linear ;
(10) $RTr_{A/K}(a) = na$ if $a \in K$ ($n^2 = |A:K|$) ;
(11) $RTr_{A/K}(aa') = RTr_{A/K}(a'a)$ ($a,a' \in A$) .

Furthermore, we find (cf. (6) and Lemma 3)

(12) $RN_{A/K}(a) \neq 0$ if and only if $a \in A^*$.

Before we proceed we need to make a remark: if D is a skew field then we may define the trace $\text{tr}: M_n(D) \to D$ in the usual way (i.e. the sum of the entries of the main diagonal), however we must handle this function with care: for instance, it will *not* in general fulfill $\text{tr}(AB) = \text{tr}(BA)$ as in the case of a commutative ring.

Theorem 1 . *Let D be a K-skew field, $A := M_n(D)$ (i.e. A a central simple K-algebra), $\delta\varepsilon\tau: A \to D$ according to* Definition 1 in §20. *and* $\text{tr}: A \to D$ *the trace, then* ($a \in A$)

$$RN_{A/K}(a) = RN_{D/K}(\delta\varepsilon\tau(a)) \quad and \quad RTr_{A/K}(a) = RTr_{D/K}(\text{tr}(a)) \quad .$$

Proof. Let E be a maximal commutative subfield of D, hence a splitting field of D *and* A (cf. Corollary 6 in §9.); this implies $i := i(D) = |E:K|$. Now denote $f: D \otimes_K E \xrightarrow{\sim} M_i(E)$ the isomorphism which exists by our assumptions on E; then we find isomorphisms

$$g: A \otimes_K E \xrightarrow{\sim} M_n(D \otimes_K E), \quad f_n: M_n(D \otimes_K E) \xrightarrow{\sim} M_n(M_i(E)) \text{ and}$$

$$h := f_n g: A \otimes_K E \xrightarrow{\sim} M_n(M_i(E)) = M_{ni}(E)$$

(g is the E-linear extension of the K-linear map $M_n(D) \to M_n(D \otimes_K E)$ which arises naturally from the embedding $D \to D \otimes_K E$, $d \mapsto d \otimes 1$; f_n arises componentwise from f). The idea of the proof is that we compute (say) the reduced norm with regard to D (resp. A) with the aid of f (resp. h); so if $a = (a_{rs}) \in M_n(D) = A$ then $h(a \otimes 1)$ is a block matrix with the (i,i)-block $(f(a_{rs} \otimes 1)) \in M_i(E)$ in the position (r,s). It follows (cf. (9))

$$RTr_{A/K}(a) = tr(h(a \otimes 1)) = \sum_{r=1}^{n} tr(f(a_{rr} \otimes 1)) = \sum_{r=1}^{n} RTr_{D/K}(a_{rr}) =$$

$$= RTr_{D/K}(\sum_{r=1}^{n} a_{rr}) = RTr_{D/K}(tr(a)) ,$$

hence the second of the formulae claimed. The first one is more sophisticated: clearly it suffices to prove it in case $a \in A^* = GL_n(D)$ only (for otherwise both sides of the equation vanish; cf. (12) and Definition 1 in §20.); then - since both functions $RN_{A/K}$ and $RN_{D/K} \delta \varepsilon \tau$ are multiplicative (cf. (7) and Theorem 1 in §20.) although $\delta \varepsilon \tau$ is *not* in general (cf. e.g. Example 1 in §20.) - we may use the trick already known from the proof of Lemma 1, namely - by Corollary 4 in §20. - we may assume that

$$a = (a_{rs}) \in GL_n(D) = A^* \text{ is triangular.}$$

Then $h(a \otimes 1)$ is block triangular with the (i,i)-blocks $(f(a_{rr} \otimes 1)) \in M_i(E)$ in the block main diagonal. Again we may use standard results from Linear Algebra concerning determinants of triangular block matrices and obtain (cf. also (7))

$$RN_{A/K}(a) = det(h(a \otimes 1)) = \prod_{r=1}^{n} det(f(a_{rr} \otimes 1)) = \prod_{r=1}^{n} RN_{D/K}(a_{rr}) =$$

$$= RN_{D/K}(\prod_{r=1}^{n} a_{rr}) = RN_{D/K}(\delta \varepsilon \tau(a)) . \quad \square$$

The preceding theorem can be restated in terms of the functor K_1 (see Definition 5 in §20.) with the aid of Wedderburn's Main Theorem in §3. (here $K_1(det)$ is the isomorphism described in Theorem 4 in §20.).

Corollary 1 . *Let* A *be a central simple K-algebra and* D *its skew field component (i.e.* $A \simeq M_n(D)$ *for some* n *), then* $RN_{A/K}$ *(resp.*

$RN_{D/K}$) *induce homomorphisms* $K_1(RN_{A/K})$
(*resp.* $K_1(RN_{D/K})$) *such that the diagram on the right hand side commutes*
(*here* $A \neq M_2(F_2)$):

$$\begin{array}{ccc} K_1(A) & \xrightarrow{K_1(RN_{A/K})} & K_1(K) = K^* \\ {\scriptstyle K_1(\det)}\downarrow & & \| \\ K_1(D) & \xrightarrow{K_1(RN_{D/K})} & K_1(K) \end{array}$$ □

Lemma 4 . *Let* $[A] \in Br(K)$, *then* $RTr_{A/K}: A \to K$ *is surjective.*

Proof. If it were not surjective it would have to vanish identically (because it is K-linear by (9)). Let E be a splitting field of A , i.e. $f: A \otimes_K E \simeq M_n(E)$, and $\{a_1,..,a_m\}$ a K-basis of A ($m = n^2$) , then $\{a_1 \otimes 1,..,a_m \otimes 1\}$ is an L-basis of $A \otimes_K L$ (cf. Theorem 3 in §4.), hence tr would consequently vanish on $M_n(E)$ (because of $tr(f(a_j \otimes 1)) = RTr_{A/K}(a_j) = 0$ for all j) which is obviously nonsense ! □

Needless to say, thanks to Lemma 3 the (ordinary) trace $Tr_{A/K}$ vanishes identically if the characteristic of K divides $|A:K|$, hence the reduced trace obviously gives more information than the trace (which gives no information at all in the above mentioned case).

Theorem 2 . *Let* $[A] \in Br(K)$ *and* L/K *a (not necessarily finite) field extension, then* ($a \in A$)

$$RN_{A \otimes_K L/L}(a \otimes 1) = RN_{A/K}(a) \quad and \quad RTr_{A \otimes_K L/L}(a \otimes 1) = RTr_{A/K}(a) \quad .$$

Proof. Let E be a splitting field of $A \otimes_K L$, hence one of A (see Lemma 2 in §5.), and let $h: A \otimes_K E \xrightarrow{\sim} (A \otimes_K L) \otimes_L E$ be the isomorphism from Lemma 2 in §5. (such that $h(a \otimes x) = (a \otimes 1) \otimes x$) , then - if g: $(A \otimes_K L) \otimes_L E \xrightarrow{\sim} M_n(E)$ is the isomorphism which exists by our assumptions - we may consider the isomorphism $f := gh: A \otimes_K E \xrightarrow{\sim} M_n(E)$. It follows

$$RN_{A \otimes_K L/L}(a \otimes 1) = \det(g((a \otimes 1) \otimes 1)) = \det(f(a \otimes 1)) = RN_{A/K}(a)$$

and similarly for the reduced trace. □

Again we want to restate the first claim of the preceding theorem in terms of K_1 ; this gives (with the notations introduced in Corollary 1)

Corollary 2 . *Let* $[A] \in Br(K)$ *and* L/K *a (not necessarily finite) field extension, then the embedding* $i_{L/K}: A \to A \otimes_K L$, $a \mapsto a \otimes 1$ *induces a homomorphism* $K_1(i_{L/K})$ *such that the diagram shown commutes (here* $A \neq M_2(F_2)$):

$$\begin{array}{ccc} K_1(A \otimes_K L) & \xrightarrow{K_1(RN_{A \otimes_K L/L})} & K_1(L) = L^* \\ {\scriptstyle K_1(i_{L/K})}\uparrow & & \uparrow{\scriptstyle \cup} \\ K_1(A) & \xrightarrow{K_1(RN_{A/K})} & K_1(K) = K^* \end{array}$$ □

Theorem 3. *Let* $[A], [B] \in Br(K)$ *and* $m^2 = |B:K|$, *then* ($a \in A$)
$$RN_{A\otimes_K B/K}(a\otimes 1) = (RN_{A/K}(a))^m \text{ and } RTr_{A\otimes_K B/K}(a\otimes 1) = m(RTr_{A/K}(a)).$$

Proof. Take a common splitting field E of A and B, and consider the isomorphisms $f: A \otimes_K E \xrightarrow{\sim} M_n(E)$ and $g: B \otimes_K E \xrightarrow{\sim} M_m(E)$ (where $n^2 = |A:K|$); now if $\phi: (A \otimes_K B) \otimes_K E \xrightarrow{\sim} (A \otimes_K E) \otimes_E (A \otimes_K E)$ is the isomorphism according to Lemma 3 in §5. (such that $\phi((a\otimes b)\otimes x) = (a\otimes 1)\otimes(b\otimes x)$) and $\psi: M_n(E) \otimes_E M_m(E) \xrightarrow{\sim} M_m(M_n(E)) = M_{mn}(E)$ is the one described in Corollary 1 in §5. (which is such that $\psi(a\otimes 1) = a1$) we consider the isomorphism $h := \psi(f\otimes g)\phi: (A \otimes_K B) \otimes_K E \xrightarrow{\sim} M_m(M_n(E)) = M_{mn}(E)$. Here it is easily seen that $h((a\otimes 1)\otimes 1)$ is a diagonal block matrix with m identical (n,n)-blocks $f(a\otimes 1) \in M_n(E)$ in the main block diagonal, hence
$$RN_{A\otimes_K B/K}(a\otimes 1) = \det(h((a\otimes 1)\otimes 1)) = (\det(f(a\otimes 1)))^m = (RN_{A/K}(a))^m$$
and similarly for the reduced trace. □

In terms of the functor K_1 the above amounts to

Corollary 3. *Let* $[A], [B] \in Br(K)$ *and* $m^2 = |B:K|$, *then the embedding* $i_{B/K}$: $A \to A \otimes_K B$, $a \mapsto a\otimes 1$ *induces a homomorphism* $K_1(i_{B/K})$ *such that the diagram shown commutes* (*here* $A \neq M_2(F_2)$):

$$\begin{array}{ccc} K_1(A\otimes_K B) & \xrightarrow{K_1(RN_{A\otimes_K B/K})} & K_1(K) \\ K_1(i_{B/K}) \uparrow & & \uparrow x \mapsto x^m \\ K_1(A) & \xrightarrow{K_1(RN_{A/K})} & K_1(K) \end{array}$$ □

Before we discuss the next result (which is more of a technical nature) we want to remind the reader of a few simple facts: if A is a ring, then any automorphism as well as any antiautomorphism of A preserves the centre $Z(A)$ (although, in general, not elementwise); see also (5) and (12) in §7. In this sense we claim

Lemma 5. *Let* $[A] \in Br(K)$ *and* Θ *a ring automorphism (resp. antiautomorphism) of* A, *then* ($a \in A$)
$$RN_{A/K}(\Theta(a)) = \Theta(RN_{A/K}(a)) \text{ and } RTr_{A/K}(\Theta(a)) = \Theta(RTr_{A/K}(a)).$$

Proof. Select a splitting field E of A such that $E \subseteq A$ (e.g. take a maximal commutative subfield of the skew field component of A; cf. Corollary 6 in §9.), hence $\Theta(E)$ makes sense and is likewise a subfield of A. Now consider the E-algebra isomorphism $f: A \otimes_K E \xrightarrow{\sim} M_n(E)$ (which exists by assumption on E) and the isomorphism (resp. antiisomorphism) $\Theta_n: M_n(E) \xrightarrow{\sim} M_n(\Theta(E))$ of rings which arises componentwise from Θ (resp. which arises componentwise from Θ via the transpose), then it is easily seen by inspection that $g := \Theta_n f(\Theta \otimes \theta |_E)^{-1} : A \otimes_K \Theta(E) \xrightarrow{\sim} M_n(\Theta(E))$ is a $\Theta(E)$-algebra isomorphism, hence $\Theta(E)$ is likewise a splitting field of A. Moreover ($a \in A$)

$$RN_{A/K}(\Theta(a)) = \det(g(\Theta(a)\otimes 1)) = \det(\Theta_n(f(a\otimes 1))) =$$
$$= \Theta(\det(f(a\otimes 1))) = \Theta(RN_{A/K}(a))$$

and similarly for the reduced trace. □

Now we come to the principal result of this paragraph:

Reduced Tower Formulae. *Let* $[A] \in Br(K)$ *and* B *a simple subring of* A *such that* $L := Z(B) \supseteq K$. *Moreover, if* $C := Z_A(L)$, *then* $Z(C) = L$, $[B],[C] \in Br(L)$, $t^2 := |Z_C(B):L|$ *and we have the formulae*

$$RN_{A/K}(b) = (N_{L/K}(RN_{B/L}(b)))^t \text{ and }$$

$$RTr_{A/K}(b) = t(Tr_{L/K}(RTr_{B/L}(b))) \text{ for all } b \in B \subseteq A.$$

Corollary 4. *Let* $[A] \in Br(K)$, $n^2 = |A:K|$ *and* L/K *a field extension of degree* r *such that* $K \subseteq L \subseteq A$, *then*

$$RN_{A/K}(x) = (N_{L/K}(x))^{n/r} \text{ and } RTr_{A/K}(x) = \frac{n}{r}(Tr_{L/K}(x))$$

for all $x \in L \subseteq A$.

Corollary 5. *Let* $[A] \in Br(K)$, L/K *a field extension such that* $K \subseteq L \subseteq A$. *If* $B := Z_A(L)$, *then* $[B] \in Br(L)$ *and we have the formulae*

$$RN_{A/K}(b) = N_{L/K}(RN_{B/L}(b)) \text{ and }$$

$$RTr_{A/K}(b) = Tr_{L/K}(RTr_{B/L}(b)) \text{ for all } b \in B \subseteq A.$$

Proof. The proof will be somewhat lengthy (for alternative proofs cf. Exercises 1/2); let us start with the corollaries: Corollary 4 is clear, because in the case $B = L$ we have $Z_C(B) = C \cap Z_A(L) = C = Z_A(L)$ and consequently $t^2 = |Z_A(L):L| = n^2/r^2$ thanks to *(vii)* in the Centralizer Theorem in §7.; Corollar 5 is also an easy consequence of the Reduced Tower Formulae because of

$$Z_C(B) = C \cap Z_A(B) = C \cap Z_A(Z_A(L)) = C \cap L = L$$

thanks to *(iv)* in the Centralizer Theorem in §7.

Now we start with the actual proof of the Reduced Tower Formulae: first we claim that

(13)   it suffices to prove Corollary 5 only.

Indeed, $B \subseteq C$ is clear from the definitions and $[B],[C] \in Br(L)$ follows from the Centralizer Theorem in §7., hence

$$C \simeq B \otimes_L Z_C(B) \text{ because of Corollary 8 in §7.}$$

Now, if we identify $b \in B \subseteq C$ with $b \otimes 1 \in B \otimes_L Z_C(B) \simeq C$ we obtain the Reduced Tower Formulae from Corollary 5 in connection with Theorem 3 ; this proves (13).

In the second step of our proof we show that

(14) it suffices to prove Corollary 5 in the cases " L/K purely inseparable" and " L/K separable" respectively.

Indeed, if M denotes the separable closure of K in L , i.e. L/M purely inseparable and M/K separable, we introduce $D := Z_B(M) =$
$= B \cap Z_A(M) = Z_A(L) \cap Z_A(M) = Z_A(M)$ (cf. (8) in §7.) which is such that [D] $\in$ Br(M) , $B \subseteq D \subseteq A$ and $Z_D(L) = D \cap Z_A(L) = D \cap B = B$ (cf. the Centralizer Theorem in §7.). It follows (with the aid of the Tower Formulae for field extensions)

$$RN_{A/K}(b) = N_{M/K}(RN_{D/M}(b)) = N_{M/K}(N_{L/M}(RN_{B/L}(b))) =$$
$$= N_{L/K}(RN_{B/L}(b)) \quad \text{for all} \quad b \in B \subseteq D \subseteq A$$

and similarly for the reduced trace.

In the third step we shall prove

(15) Corollary 5 is true in case " L/K purely inseparable".

For the proof of (15) let E be a splitting field of A such that (say) E/K is finite Galois (cf. Theorem 9 in §9.); this implies $L \cap E = K$ and $L \otimes_K E$ is a field $\simeq$ LE (note that the fact that $L \otimes_K E$ is a field is the only fact that matters in our context), and from Field Theory we know then

(16) $N_{LE/E}(x) = N_{L/K}(x)$ and $Tr_{LE/E}(x) = Tr_{L/K}(x)$ for all $x \in L$.

By virtue of

$[B] = [A \otimes_K L]$ in Br(L) (cf. Corollary 1 in §9.)

we see that LE is a splitting field of B (cf. Lemma 2 in §5.), hence we have an LE-algebra (and therefore an E-algebra) isomorphism g: $B \otimes_L LE \xrightarrow{\sim} M_s(LE)$ (here $s := n/|L:K|$ where $n^2 = |A:K|$) in addition to the (assumed) E-algebra isomorphism f: $A \otimes_K E \xrightarrow{\sim} M_n(E)$. Now consider the maps (here h is according to Lemma 2 in §5.)

$$\begin{array}{ccccc} B \otimes_L (L \otimes_K E) & \simeq & B \otimes_L LE & \xrightarrow{g} & M_s(LE) \simeq End_{LE}(LE^s) \\ \phi \uparrow & & & & \cap \\ B \otimes_K E & & \xrightarrow{\quad h \quad} & & M_n(E) \simeq End_E(LE^s) \quad ; \end{array}$$

clearly h is an E-algebra homomorphism and we find from (16) and Lemmas 1/2 the equation

$$N_{L/K}(RN_{B/L}(b)) = N_{LE/E}(det_{LE}(g(b \otimes 1))) = det_E(g(b \otimes 1)) =$$

$$= \det(h(b\otimes 1)) \quad \text{for all} \quad b \in B \ .$$

On the other hand the embedding $\psi: B \hookrightarrow A$ gives an E-algebra homomorphism $h' := f(\psi \otimes \mathrm{id}_E): B \otimes_K E \to M_n(E)$ such that

$$RN_{A/K}(b) = \det(h'(b\otimes 1)) \quad \text{for all} \quad b \in B \ ;$$

now since $B \otimes_K E$ is simple by Theorem 5 in §5. (for it is $\simeq B \otimes_L LE$ as we have seen above) the Skolem-Noether Theorem in §7. implies

$$h'(b\otimes 1) = th(b\otimes 1)t^{-1} \quad \text{for some} \quad t \in GL_n(E) \quad (b \in B) \ ,$$

hence (15) since the reduced traces can be handled in the same way.

Now we come to the fourth step: we want to prove

(17)   Corollary 5 is true in case " L/K separable".

Again select a splitting field E of A such that E/K is finite Galois (cf. Theorem 9 in §9.); E is then likewise a splitting field of B (thanks to $[B] = [A \otimes_K L]$ in $Br(L)$; cf. Corollary 1 in §9.). Of course we may assume $L \subseteq E$; now we claim that

(18)   it suffices to assume $E \subseteq A$

for otherwise we replace A by $A' := M_t(A)$ where t is suitable such that $E \subseteq A'$ (cf. Lemma 5 in §9.; note $[A] = [A'] \in Br(K)$). Then $B' := Z_{A'}(L) = M_t(B)$ (see e.g. the Centralizer Theorem in §7.: "$\supseteq$" is true trivially and "$=$" follows then for dimensional reasons) and E is clearly a splitting field of both A' and B'. Now let $b \in B$ be given; set

$$b' := \begin{pmatrix} 1 & & 0 \\ & \ddots_1 & \\ 0 & & b \end{pmatrix} \quad (\text{resp.} := \begin{pmatrix} 0 & & 0 \\ & \ddots_0 & \\ 0 & & b \end{pmatrix}) \in B' \subseteq A' \ ,$$

then (cf. Theorem 1) $RN_{A'/K}(b') = RN_{A/K}(b)$ and $RN_{B'/L}(b') = RN_{B/L}(b)$ (resp. $RTr_{A'/K}(b') = RTr_{A/K}(b)$ and $RTr_{B'/L}(b') = RTr_{B/L}(b)$), hence (18). So the fifth (and final) step in our proof will consist of showing

(19)   Corollary 5 is true in case there exists a splitting field E of A such that E/K is Galois and $K \subseteq L \subseteq E \subseteq A$.

Our proof of (19) will be lengthy (but entirely straightforward in some sense): consider $C := Z_B(E) = B \cap Z_A(E) = Z_A(L) \cap Z_A(E) = Z_A(E) \subseteq B$, denote $\Gamma := \mathrm{Gal}(E/K) \geq \mathrm{Gal}(E/L) =: \Delta$ and choose a system R of representatives for the cosets of Γ modulo Δ, i.e. $\Gamma = \bigcup_{\rho \in R} \Delta\rho = \bigcup_{\rho \in R} \rho^{-1}\Delta$, hence (use Theorem 5 in §7. twice (for the notation cf. its proof))

(20)
$$B = \bigoplus_{\delta \in \Delta} e_\delta C = \bigoplus_{\delta \in \Delta} Ce_\delta \quad \text{and}$$

$$A = \bigoplus_{\rho \in R}\bigoplus_{\delta \in \Delta} e_\delta e_\rho C = \bigoplus_{\rho \in R}\bigoplus_{\delta \in \Delta} Ce_\delta e_\rho = \bigoplus_{\rho \in R} Be_\rho \ ;$$

in (20) we have in general $B \neq e_\rho B e_\rho^{-1}$ but always $e_\rho B e_\rho^{-1} \subseteq C$.

Moreover, by construction E is a common splitting field of A, B and C (cf. Corollary 1 in §9.) hence

$$f: C \xrightarrow{\sim} M_m(E) \text{ as E-algebras for some } m.$$

Now fix $\rho \in R$ and define another E-algebra isomorphism

$$^\rho f: C \xrightarrow{\sim} M_m(E) , \quad c \mapsto {}^\rho(f(e_\rho^{-1} c e_\rho))$$

where ρ acts on the matrices on the right-hand side componentwise (cf. the proof of Theorem 5 in §7.). The Skolem-Noether Theorem in §7. then gives ($\rho \in R$, $c \in C$)

(21) $\qquad {}^\rho f(c) = t_\rho f(c) t_\rho^{-1}$ for some $t_\rho \in GL_m(E)$.

Now regard B via (20) as a free right C-module of rank $h := |\Delta| = |E:L|$; if we denote $L: B \to \text{End}_C(B)$, $b \mapsto L_b$ the L-algebra homomorphism arising from left multiplication we obtain the isomorphisms of E-algebras

$$\begin{array}{ccc} \text{End}_C(B) \simeq M_h(C) & \xrightarrow{f_h} & M_h(M_m(E)) \\ L_E \Big\uparrow \wr & & \Big\| \\ B \otimes_L E & \xrightarrow{g} & M_s(E) \end{array} \qquad (s := hm)$$

where f_h arises from f componentwise and L_E stands for the E-linear extension of L (cf. Theorem 3 in §5.) which is an isomorphism for dimensional reasons (note that it is injective since $B \otimes_L E$ has no proper two-sided ideals thanks to Corollary 3 in §5.). We find (cf. (20))

(22) $\qquad g(b\otimes 1) = (f(c_{\varepsilon\delta}))$ where $b e_\delta = \sum_{\varepsilon \in \Delta} e_\varepsilon c_{\varepsilon\delta}$ ($\delta \in \Delta$).

Clearly we have for all $b \in B$

(23) $\qquad \begin{aligned} RN_{B/L}(b) &= \det(g(b\otimes 1)) = \det(f(c_{\varepsilon\delta})) \text{ and} \\ RTr_{B/L}(b) &= \text{tr}(g(b\otimes 1)) = \sum_{\delta \in \Delta} \text{tr}(f(c_{\delta\delta})) \text{ (here tr in } M_m(E) \text{)}. \end{aligned}$

On the other hand - again via (20) - we may now regard A as a free right C-module of rank $rh = |\Gamma| = |E:K|$ ($r := |L:K| = |R|$) and consider the E-algebra isomorphisms (argue as above)

$$\begin{array}{ccc} \text{End}_C(A) \simeq M_{rh}(E) & \xrightarrow{f_{rh}} & M_{rh}(M_m(E)) \\ L_E \Big\uparrow \wr & & \Big\| \\ A \otimes_K E & \xrightarrow{k} & M_n(E) \end{array} \qquad (n := rhm , n^2 = |A:K|)$$

which yield the formulae

$$RN_{A/K}(a) = \det(k(a\otimes 1)) \text{ and } RTr_{A/K}(a) = \text{tr}(k(a\otimes 1)) \quad (a \in A).$$

Now assume $b \in B \subseteq A$; it follows (cf. (22))

$$L_b(e_\delta e_\rho) = be_\delta e_\rho = \sum_{\varepsilon \in \Delta} e_\varepsilon c_{\varepsilon\delta} e_\rho = \sum_{\varepsilon \in \Delta} e_\varepsilon e_\rho (e_\rho^{-1} c_{\varepsilon\delta} e_\rho) \quad (b \in B),$$

hence - if $t := \begin{pmatrix} \rho^{-1} & 0 \\ 0 & t_\rho \end{pmatrix} \in M_r(M_s(E))^* = GL_n(E)$ (cf. (21)) - thanks to (21) and (22):

$$k(b \otimes 1) = \begin{pmatrix} f(e_\rho^{-1} c_{\varepsilon\delta} e_\rho) & 0 \\ 0 & \end{pmatrix} = \begin{pmatrix} \rho^{-1}(t_\rho f(c_{\varepsilon\delta}) t_\rho^{-1}) & 0 \\ 0 & \end{pmatrix} =$$

(24)
$$= t \begin{pmatrix} \rho^{-1}(f(c_{\varepsilon\delta})) & 0 \\ 0 & \end{pmatrix} t^{-1} = t \begin{pmatrix} \rho^{-1}(g(b \otimes 1)) & 0 \\ 0 & \end{pmatrix} t^{-1} \in M_r(M_s(E))$$

for all $b \in B \subseteq A$.

Using standard results from Linear Algebra concerning determinants and traces of diagonal block matrices we obtain immediately from (23) and (24) the formulae

$$RN_{A/K}(b) = \det(k(b \otimes 1)) = \prod_{\rho \in R} \rho^{-1}(RN_{B/L}(b)) = N_{L/K}(RN_{B/L}(b))$$

for all $b \in B \subseteq A$

and similarly for the reduced trace. □

Exercise 1. Study section 9 (pp.112) of I. Reiner [1975] where you can find a different approach to reduced norms and traces.

Exercise 2. Give an alternative proof for the Reduced Tower Formulae by showing that a specialisation argument (involving rational function fields in one variable) reduces the problem to the much simpler (ordinary) Tower Formulae (see P. Draxl & M. Kneser [1980]).

Exercise 3. Let H be the skew field of real quaternions (see §1.); describe the function $RN_{H/R}$ explicitly and show

$$RN_{H/R}(H^*) = R^*_{>0} := \text{multiplicative group of positive reals}.$$

For another interesting result cf. W.C. Waterhouse [1982]

§ 23. THE REDUCED WHITEHEAD GROUP $SK_1(D)$ AND WANG'S THEOREM

In this paragraph we want to study the Whitehead group $K_1(D)$ $:= D^{*ab}$ (cf. Definition 5 in §20.) of a K-skew field D by comparing it with $K_1(K) = K^*$. Thanks to Wedderburn's Theorem in §10. we may restrict ourself to the case of *infinite* base fields K.

If A is a central simple K-algebra (e.g. a K-skew field), then $K_1(A)$ and $K_1(K) = K^*$ are linked by the homomorphism $K_1(RN_{A/K})$ which is induced by the *reduced norm* (cf. Corollary 1 in §22.). In this sense we have

Definition 1. *Let* $[A] \in Br(K)$, *then*

$$SK_1(A) := \operatorname{Ker} K_1(RN_{A/K}) = \{ a \in A \mid RN_{A/K}(a) = 1 \}/[A^*,A^*]$$

is called the "reduced Whitehead group of A *" and*

$$SH^0(A) := \operatorname{Coker} K_1(RN_{A/K}) = K^*/RN_{A/K}(A^*)$$

the "reduced norm residue group of A *"*.

Obviously we get

(1) $\qquad SK_1(K) = 1$ and $SH^0(K) = 1$

as well as (cf. Lemma 5 in §22.)

(2) $\qquad SK_1(A^{op}) = SK_1(A)$ and $SH^0(A^{op}) = SH^0(A)$.

Now denote D the skew field component of A (i.e. $A \simeq M_n(D)$ for some n); thanks to Corollary 1 in §22. we have a commutative diagram

(3)
$$\begin{array}{ccccccccc}
1 & \to & SK_1(A) & \to & K_1(A) & \to & K_1(K) & \to & SH^0(A) & \to & 1 \\
& & \downarrow \wr & & \downarrow K_1(\det) \wr & & \| & & \| & & \\
1 & \to & SK_1(D) & \to & K_1(D) & \to & K_1(K) & \to & SH^0(D) & \to & 1
\end{array}$$

where the rows are exact and $K_1(\det)$ is as in Theorem 4 in §20. Of course (3) means that $SK_1(A)$ and $SH^0(A)$ depend (up to isomorphism) on the class [A] in the Brauer group Br(K) only.

Now let us reinterpret Corollary 2 in §22. in terms of Definition 1 : it says that we have a commutative diagram

(4)
$$\begin{array}{ccccccccc} 1 & \longrightarrow & SK_1(A\otimes_K L) & \longrightarrow & K_1(A\otimes_K L) & \longrightarrow & K_1(L) & \longrightarrow & SH^0(A\otimes_K L) & \longrightarrow & 1 \\ & & \uparrow & & \uparrow K_1(i_{L/K}) & & \uparrow U & & \uparrow & & \\ 1 & \longrightarrow & SK_1(A) & \longrightarrow & K_1(A) & \longrightarrow & K_1(K) & \longrightarrow & SH^0(A) & \longrightarrow & 1 \end{array}$$

with exact rows (for any field extension L/K).

Corollary 2 in §22. has a converse:

Lemma 1. *Let* [A] ∈ Br(K) *and* L/K *a finite field extension of degree* m := |L:K| ; *then there is a homomorphism* $K_1(N_{L/K})$ *such that the diagram shown commutes and such that*

$$K_1(N_{L/K})K_1(i_{L/K}) =$$
$$= |L:K| id_{K_1(A)} .$$

$$\begin{array}{ccc} K_1(A\otimes_K L) & \xrightarrow{K_1(RN_{A\otimes_K L/L})} & K_1(L) = L^* \\ K_1(N_{L/K}) \downarrow & & \downarrow N_{L/K} \\ K_1(A) & \xrightarrow{K_1(RN_{A/K})} & K_1(K) = K^* \end{array}$$

Proof. Let $L: L \to M_m(K)$ be the K-algebra homomorphism arising from left multiplication together with the standard isomorphism $End_K(L) \simeq M_m(K)$; then $id_A \otimes L$ induces a group homomorphism (cf. Lemma 4 in §5.)

$$\pi: K_1(A\otimes_K L) \longrightarrow K_1(A\otimes_K M_m(K)) \simeq K_1(M_m(A)) .$$

Now we see from Theorem 2 in §7. that the embedding $A \otimes_K L \hookrightarrow A \otimes_K M_m(K)$ is such that $Z_{A\otimes_K M_m(K)}(K \otimes_K L) = Z_A(K) \otimes_K Z_{M_m(K)}(L) = A \otimes_K L$ (note that $Z_{M_m(K)}(L) = L$ follows for dimensional reasons from the Centralizer Theorem in §7. if we view L as embedded into $M_m(K)$ via L above), hence

(5) $RN_{A\otimes_K M_m(K)/K}(x) = N_{L/K}(RN_{A\otimes_K L/L}(x))$ for all $x \in A \otimes_K L$

by the reduced Tower Formulae in §22. (or Corollary 5 in §22.).

Now assume A a K-skew field (we may do this in view of (3) above) and define

$$K_1(N_{L/K}) := K_1(det)\pi \text{ with } det: GL_m(A) \to A^{*ab} = K_1(A) .$$

Now (5) and Corollary 1 in §22. imply the commutativity of the diagram shown in our lemma, hence we are left with the proof of the relation $K_1(N_{L/K})K_1(i_{L/K}) = |L:K| id_{K_1(A)}$; for this we consider the two K-algebra homomorphisms

$$f,g: A \to M_m(A) \simeq A \otimes_K M_m(K) \; ; \; f(a) = a \otimes L_1 \; , \; g(a) = a1$$

which differ by an inner automorphism of $M_m(A)$ only (cf. the Skolem-Noether Theorem in §7.). The latter implies that the classes $\overline{f(a)}$ and $\overline{g(a)}$ coincide in the factor group $K_1(M_m(A)) = GL_m(A)^{ab}$, hence - thanks to $\delta\epsilon\tau(a1) = a^m$ (see §20.) - we obtain

$$K_1(N_{L/K})K_1(i_{L/K})(\overline{a}) = K_1(N_{L/K})(\overline{a \otimes 1}) = \det(\overline{a \otimes L_1}) = \det(\overline{a1}) =$$
$$= \overline{\delta\epsilon\tau(a1)} = \overline{a}^m = \overline{a}^m = |L:K| id_{K_1(A)}(\overline{a})$$

for all $\overline{a} \in K_1(A)$ ($a \in A^*$). □

An immediate consequence of Lemma 1 is the existence and commutativity of the diagram

(6)
$$\begin{array}{ccccccccc} 1 & \to & SK_1(A \otimes_K L) & \to & K_1(A \otimes_K L) & \to & K_1(L) & \to & SH^0(A \otimes_K L) & \to & 1 \\ & & \downarrow & & {\scriptstyle K_1(N_{L/K})}\downarrow & & \downarrow{\scriptstyle N_{L/K}} & & \downarrow & & \\ 1 & \to & SK_1(A) & \to & K_1(A) & \to & K_1(K) & \to & SH^0(A) & \to & 1 \end{array}$$

with exact rows (for any finite field extension L/K).
Since there is always a splitting field L of A such that $|L:K| = i(A)$ we obtain from (1),(3),(4),(6) and the last assertion of Lemma 1 immediately the important

Lemma 2. *Let* $[A] \in Br(K)$, *then* $SK_1(A)$ *and* $SH^0(A)$ *are abelian torsion groups with fixed exponent dividing* $i(A)$. □

Later in this paragraph we shall strengthen the first of the two assertions of the preceding lemma.

Another immediate consequence of the last assertion of Lemma 1 (together with Lemma 2 is)

Lemma 3. *Let* $[A] \in Br(K)$ *and* L/K *a finite field extension of degree* $m = |L:K|$; *then* $a \mapsto a \otimes 1$ *(resp. the embedding* $K \subseteq L$ *) induce injections* $SK_1(A) \hookrightarrow SK_1(A \otimes_K L)$ *(resp.* $SH^0(A) \hookrightarrow SH^0(A \otimes_K L)$ *) provided* $|L:K|$ *and* $i(A)$ *are coprime*. □

The first assertion of Lemma 2 may be stated in a more *quantitative* way:

Lemma 4. *Let* D *be a K-skew field of index* $i := i(D)$; *if* $d \in D^*$ *such that* $RN_{D/K}(d) = 1$ *(for instance if* $d \in [D^*, D^*]$ *) then* d *is a product of at most* $2(i^2-1)$ *commutators of elements in* D^* .

Proof. Take a splitting field L of D (which implies the existence of a K-algebra isomorphism $f: D \otimes_K L \xrightarrow{\sim} M_i(L)$) and consider the two K-algebra homomorphisms $g, h: D \to M_i(D)$ defined by $g(d) := d1$ and

$$
\begin{array}{ccc}
d \otimes 1 & D \otimes_K L \xrightarrow{\sim}_{f} M_i(L) \\
\uparrow & \uparrow & \downarrow \\
d & D \xrightarrow[h]{- - - -} M_i(D)
\end{array}
$$

The Skolem-Noether Theorem implies $g(d) = th(d)t^{-1}$ for some $t \in GL_i(D)$; by definition of the reduced norm in §22. (together with the fact that $\delta\epsilon\tau$ coincides with the usual determinant in the commutative case; cf. Theorem 1/Corollary 1 in §20.) we get

$$\delta\epsilon\tau(h(d)) = \delta\epsilon\tau(f(d \otimes 1)) = RN_{D/K}(d) = 1 \quad ,$$

hence

$$d^i = \delta\epsilon\tau(d1) = \delta\epsilon\tau(g(d)) = \delta\epsilon\tau(th(d)t^{-1}) =$$
$$= \text{a product of at most } 2(i^2 - 1) \text{ commutators of elements in } D^*$$

because of Corollary 2 in §20. □

Now let us reinterpret Corollary 3 in §22. in terms of Definition 1 : it says that for $[A], [B] \in Br(K)$ we have a commutative diagram

(7)
$$
\begin{array}{ccccccccc}
1 & \to & SK_1(A \otimes_K B) & \to & K_1(A \otimes_K B) & \to & K_1(K) & \to & SH^0(A \otimes_K B) & \to & 1 \\
& & \uparrow & & \uparrow K_1(i_{B/K}) & & \uparrow \times i(B) & & \uparrow \\
& & & & & & \uparrow \times & & \\
1 & \to & SK_1(A) & \to & K_1(A) & \to & K_1(K) & \to & SH^0(A) & \to & 1
\end{array}
$$

with exact rows (here B may assumed to be a K-skew field thanks to (3), hence $i(B) = m$ (in the sense of Corollary 3 in §22.)).
Corollary 3 in §22. has also a converse:

Lemma 5. *Let* $[A], [B] \in Br(K)$, *then there is a homomorphism* $K_1(p_{B/K})$ *such that the diagram shown commutes and such that*

$$
\begin{array}{ccc}
K_1(A \otimes_K B) & \xrightarrow{K_1(RN_{A \otimes_K B/K})} & K_1(K) = K^* \\
K_1(p_{B/K}) \uparrow & & \downarrow \\
K_1(A) & \xrightarrow{K_1(RN_{A/K})} & K_1(K) = K^* \quad \times^{i(B)}
\end{array}
$$

$$K_1(p'_{B/K}) K_1(i_{B/K}) = i(B)^2 \, id_{K_1(A)} \quad .$$

Proof. It suffices to study the case where A and B are skew fields (see (3)); consider

$$K_1(i_{B^{op}/K}): K_1(A \otimes_K B) \to K_1(A \otimes_K B \otimes_K B^{op}) \simeq K_1(M_{i(B)^2}(A))$$

(remember $A \otimes_K B \otimes_K B^{op} \simeq A \otimes_K M_r(K) \simeq M_r(A)$ thanks to Lemma 3 in §4. and Corollary 2/Lemma 4 in §5.) and define

$$K_1(p_{B/K}) := K_1(\det)K_1(i_{B^{op}/K}) \text{ with } \det: GL_r(A) \to K_1(A)$$

where $r := |B:K| = i(B)^2$ (because B is a skew field). In view of $i(B^{op}) = i(B)$ the commutativity of the diagram in our lemma is clear from (3) and (7); therefore we are left with the proof of the relation $K_1(p_{B/K})K_1(i_{B/K}) = i(B)^2 \mathrm{id}_{K_1(A)}$; for this we consider the two K-algebra homomorphisms ($r := i(B)^2$ as above)

$$f,g: A \to M_r(A) \simeq A \otimes_K B \otimes_K B^{op} \ ; \ f(a) = a \otimes 1 \otimes 1 \ , \ g(a) = a1$$

which differ by an inner automorphism of $M_r(A)$ only (cf. the Skolem-Noether Theorem in §7.). The latter implies that the classes $\overline{f(a)}$ and $\overline{g(a)}$ in the factor group $K_1(M_r(A)) = GL_r(A)^{ab}$ coincide , hence - thanks to $\delta\varepsilon\tau(a1) = a^r$ (see §20.) - we obtain

$$K_1(p_{B/K})K_1(i_{B/K})(\overline{a}) = K_1(p_{B/K})(\overline{a \otimes 1}) = \det(\overline{a \otimes 1 \otimes 1}) = \det(\overline{a1}) =$$
$$= \overline{\delta\varepsilon\tau(a1)} = \overline{a^r} = \overline{a}^r = i(B)^2 \mathrm{id}_{K_1(A)}(\overline{a})$$

for all $\overline{a} \in K_1(A)$ ($a \in A^*$) . □

An immediate consequence of Lemma 5 is the existence and commutativity of the diagram

(8)
$$\begin{CD}
1 @>>> SK_1(A \otimes_K B) @>>> K_1(A \otimes_K B) @>>> K_1(K) @>>> SH^0(A \otimes_K B) @>>> 1 \\
@. @VVV @VV{K_1(p_{B/K})}V @VV{\times i(B)}V @VV{\times i(B)}V @. \\
1 @>>> SK_1(A) @>>> K_1(A) @>>> K_1(K) @>>> SH^0(A) @>>> 1
\end{CD}$$

with exact rows.

Thanks to Lemma 2,(7),(8) and the last assertion of Lemma 5 we may conclude easily:

(9)   If $i(A)$ and $i(B)$ are coprime, then there are injections
$SK_1(A) \hookrightarrow SK_1(A \otimes_K B)$ (resp. $SH^0(A) \hookrightarrow SH^0(A \otimes_K B)$) induced by $a \mapsto a \otimes 1$ (resp. $x \mapsto x^{i(B)}$) .

The statement (9) can easily be improved as follows: use (9) again (but restrict it to the $i(A)$-torsion component of $SK_1(A \otimes_K B)$ and $SH^0(A \otimes_K B)$) with $A \otimes_K B$ (resp. B^{op}) in place of A (resp. B) and apply the isomorphism $A \otimes_K B \otimes_K B^{op} \simeq M_r(A)$ ($r := i(B)^2$) which we have used already above; now observe that $K_1(p_{B/K})$ and $K_1(i_{B^{op}/K})$ differ by an automorphism only (namely by $K_1(\det)$; see the definition of the map $K_1(p_{B/K})$ in the course of the proof of Lemma 5), hence $K_1(p_{B/K})$ induces an isomorphism from the i(A)-torsion component of $SK_1(A \otimes_K B)$ onto $SK_1(A)$. Therefore $K_1(i_{B/K})$ is an isomorphism of $SK_1(A)$ onto the i(A)-torsion component of $SK_1(A \otimes_K B)$. By a similar (but actually easier) argument we see that $x \mapsto x^r$ induces an isomorphism of $SH^0(A)$ onto the i(A)-torsion component of $SH^0(A \otimes_K B)$. Interchanging the roles of A and B gives (cf. the Theory of Torsion Groups)

Lemma 6. *Let* [A],[B] $\in$ Br(K) *such that the indices* i(A) *and* i(B) *are coprime, then we have isomorphisms*

$$SK_1(A \otimes_K B) \simeq SK_1(A) \times SK_1(B) \text{ and } SH^0(A \otimes_K B) \simeq SH^0(A) \times SH^0(B). \; \square$$

In view of Corollary 11 in §9. all this implies (together with (3))

Corollary 1. *Let* [A] $\in$ Br(K), *then for the study of the structure of the groups* $SK_1(A)$ *and* $SH^0(A)$ *it suffices to consider the case of a K-skew field* A *of prime power index*. $\square$

Now let us turn to more special results (note that so far in this paragraph we discussed the general functorial behaviour of the functors SK_1 and SH^0 without restrictions).

Lemma 7. *Let* D *be a K-skew field and* L *a maximal commutative subfield of* D *such that* L/K *is either purely inseparable or cyclic. If* d $\in$ L* $\subseteq$ D* *and* $RN_{D/K}(d) = 1$, *then* d *is a commutator.*

Proof. Let us start with the *purely inseparable* case: then i(D) = |L:K| = $= p^f$ where $p := \text{char}(K) \neq 0$. Thanks to Corollary 4 in §22. we get

$$0 = RN_{D/K}(d) - 1 = N_{L/K}(d) - 1 = d^{p^f} - 1^{p^f} = (d-1)^{p^f},$$

hence d = 1 and therefore even more so our assertion (for the last two steps in the above formula cf. Field Theory). Now the *cyclic* case: let $\Gamma := \text{Gal}(L/K) = \langle \sigma \rangle$, then $N_{L/K}(d) = RN_{D/K}(d) = 1$ thanks to Corollary 4 in §22. (note i(D) = |L:K| since L is maximal commutative in D ; cf. Theorem 4 in §7.); therefore we may find some m $\in$ L* $\subseteq$ D* such that $d = {}^{\sigma}m \, m^{-1}$ (cf. Hilbert's "Satz 90" in §6.) . The Skolem-Noether Theorem in §7. implies the existence of some t $\in$ D* such that $m = tmt^{-1}$, hence $d = tmt^{-1}m^{-1}$. $\square$

Now take any quaternion skew field D - i.e. i(D) = 2 -; then, if K denotes the centre of D , every element d $\in$ D lies in some quadratic field extension L/K which must be either purely inseparable or cyclic.

Consequently the following result is clear:

Theorem 1. *Let* D *be a* K-*skew field of index* 2 *(i.e. a quaternion skew field, e.g.* $D = H$: *see* §1.) , *then* $SK_1(D) = \{1\}$, *more precisely: any* $d \in D^*$ *such that* $RN_{D/K}(d) = 1$ *is a commutator, in particular:* $[D^*, D^*]$ *consists only of commutators.* □

Before we proceed with our investigations of SK_1 we must introduce some new notions from Galois Cohomology. We start with (cf. §6.)

Lemma 8. *Let* M *be a left* Γ-*module, then*
$$I_\Gamma(M) := \{ \sum_{\sigma \in \Gamma}(^\sigma m_\sigma - m_\sigma) \mid m_\sigma \in M, \ m_\sigma = 0 \ \text{for almost all } \sigma \}$$
is a Γ-*submodule of* M; *moreover, if* Γ *is finite then* $I_\Gamma(M) \subseteq \text{Ker } N_\Gamma$.

Proof. Straightforward calculations. □

Note that $M_\Gamma := M/I_\Gamma(M)$ is the largest factor module of M on which Γ acts trivially.

Definition 2. *Let* Γ *be a finite group and* M *a left* Γ-*module, then* $H^{-1}(\Gamma, M) := \text{Ker } N_\Gamma / I_\Gamma(M)$ *is called the " (-1)-st Cohomology Group of* M *".*

Lemma 9. *Let* Γ *be a finite group and* M *a left* Γ-*module, then* $H^{-1}(\Gamma, M)$ *is an abelian torsion group with fixed exponent dividing* $|\Gamma|$.

Proof. Write $n := |\Gamma|$ and take $x \in \text{Ker } N_\Gamma$, i.e. $\sum_{\sigma \in \Gamma} {}^\sigma x = N_\Gamma(x) = 0$; it follows $nx = \sum_{\sigma \in \Gamma} {}^\sigma x - nx = \sum_{\sigma \in \Gamma}({}^\sigma x - x) \in I_\Gamma(M)$. □

Lemma 10. *Let* $\Gamma = \langle \sigma \rangle$ *be a finite cyclic group and* M *a left* Γ-*module, then* $I_\Gamma(M) = \{ {}^\sigma m - m \mid m \in M \}$.

Proof. "$\supseteq$" is clear; as for the converse we find
$$\sigma^i m - m = \sigma(\sum_{j=0}^{i-1} \sigma^j m) - (\sum_{j=0}^{i-1} \sigma^j m) \ . \ \square$$

Combination of Lemma 10 with Exercise 1 in §6. yields

(10) $\qquad H^1(\Gamma, M) \simeq H^{-1}(\Gamma, M)$ in case Γ is finite cyclic.

Note that Hilbert's "Satz 90" in §6. may now be stated in the form

(11) $\qquad H^{-1}(\Gamma, L^*) = \{1\}$ if L/K is a finite cyclic field extension with Galois group Γ .

Note that (10) stresses the fact that Hilbert's "Satz 90" is merely a consequence of Noether's Equations (see also §6.).

In what follows let L/K be a finite Galois extension with $\Gamma := \text{Gal}(L/K)$; then $\text{Br}(L)$ is a left Γ-module (cf. §9.) and we recall (cf. Lemma 2 in §8.):

(12) $\qquad [A] \in \text{Br}(L)^\Gamma$ if and only if every $\sigma \in \Gamma$ can be extended to a ring automorphism of A.

In this context we claim

Lemma 11. *Let L/K be finite Galois, $\Gamma := \text{Gal}(L/K)$ and $[A] \in \text{Br}(L)^\Gamma$; then $K_1(A)$ can be given in a natural way the structure of a left Γ-module such that $K_1(RN_{A/L})$ is a left Γ-homomorphism ($K_1(L) = L^*$ being viewed as a Γ-module in the usual way); in particular:* $\text{Im } K_1(RN_{A/L}) =$
$= RN_{A/L}(A^*)$ *is a Γ-submodule of* $L^* = K_1(L)$.

Proof. Let $\sigma \in \Gamma$ be given and ϕ_σ a ring automorphism such that $\phi_\sigma|_L = \sigma$ (cf. (12)). If ψ_σ is another such automorphism then the Skolem-Noether Theorem in §7. implies the existence of some $t \in A^*$ such that $\psi_\sigma^{-1}\phi_\sigma(a) = tat^{-1}$ for all $a \in A$. It follows that for given $\bar{a} \in K_1(A) = A^{*ab}$ ($a \in A^*$) the element

$$\overline{\sigma a} := \overline{\phi_\sigma(a)} = \overline{\psi_\sigma(tat^{-1})} = \overline{\psi_\sigma(t)\psi_\sigma(a)\psi_\sigma(t)^{-1}} = \overline{\psi_\sigma(a)} =$$
$$= \overline{\psi_\sigma(a)} \in K_1(A)$$

is independent of the choice of the extension ϕ_σ of σ, hence $\overline{\sigma a}$ is well-defined and furnishes $K_1(A)$ with the structure of a Γ-module. The rest is then clear thanks to Lemma 5 in §22. □

Now we are fully prepared for a generalization of the *cyclic* part of Lemma 7 above.

Theorem 2. *Let $[A] \in \text{Br}(K)$, L/K Galois, $L \subseteq A$, $\Gamma := \text{Gal}(L/K)$ and $B := Z_A(L)$; then $[B] \in \text{Br}(L)^\Gamma$. Moreover, denote*

$$SK_1(A)|_B := \{ b \in B^* \subseteq A^* \mid RN_{A/K}(b) = 1 \}/(B^* \cap [A^*, A^*])$$

the subgroup of $SK_1(A)$ which is generated by the elements of B^*, then we have the two exact sequences

$$SK_1(B) \xrightarrow{f} SK_1(A)|_B \xrightarrow{g} X \longrightarrow 1$$

and

$$H^{-1}(\Gamma, RN_{B/L}(B^*)) \xrightarrow{h} X \longrightarrow 1 \ .$$

Here $X := RN_{B/L}(\{b \in B^* \mid RN_{A/K}(b) = 1\})/RN_{B/L}(B^* \cap [A^*, A^*])$, f and h are induced by embeddings and g is induced by $RN_{B/L}$.

Corollary 2. *In the situation and with the notation of* Theorem 2 *we have* $SK_1(A)\big|_B = \{1\}$ *provided* $SK_1(B) = \{1\}$ *and* $H^{-1}(\Gamma, RN_{B/L}(B^*)) = \{1\}$. □

Corollary 3. *In the situation and with the notation of* Theorem 2 *it follows that the exponent of (the torsion group)* $SK_1(A)\big|_B$ *divides* $|L:K|$*-times the exponent of* $SK_1(B)$. □

Moreover, we want to point out that we may recover the cyclic part of Lemma 7 with the aid of Corollary 2 because of

$B = L$ (since L is maximal commutative) and hence
$SK_1(B) = \{1\}$ by (1)

and

$$H^{-1}(\Gamma, RN_{B/L}(B^*)) = H^{-1}(\Gamma, L^*) = \{1\} \text{ by (11)}.$$

Proof (of Theorem 2). Thanks to (6) and Corollary 1 in §9. $[B] \in Br(L)^\Gamma$ is clear and consequently $RN_{B/L}(B^*)$ is a Γ-submodule of L^*. Because of Corollary 5 in §22. f and g are well-defined and the upper sequence is clearly exact. Again by Corollary 5 in §22. we find

$$RN_{B/L}(\{b \in B^* | RN_{A/K}(b) = 1\}) = \text{Ker } N_{L/K}\big|_{RN_{B/L}(B^*)}$$

which implies that in order to prove the exactness of the lower sequence it suffices to show that h is well-defined which clearly amounts to showing the inclusion (cf. Definition 2)

$$I_\Gamma(RN_{B/L}(B^*)) \subseteq RN_{B/L}(B^* \cap [A^*, A^*]) ;$$

the latter, however, is not difficult: let $\sigma \in \Gamma$ be given, then the Skolem-Noether Theorem in §7. implies the existence of $e_\sigma \in A^*$ such that $e_\sigma B e_\sigma^{-1} = B$ and $e_\sigma x e_\sigma^{-1} = x$ for all $x \in L$ (see Theorem 5 in §7. and the proof thereof). Now use Lemma 5 in §22. for obtaining

$$^\sigma RN_{B/L}(b) RN_{B/L}(b)^{-1} = RN_{B/L}(e_\sigma b e_\sigma^{-1} b^{-1}) \in RN_{B/L}(B^* \cap [A^*, A^*])$$

for all $b \in B^*$. □

Theorem 2 (more precisely: Corollary 2) suggests the following definition which is feasible because of Lemma 11 :

Definition 3. *A commutative field* k *is called "reasonable" if for any finite separable field extension* K/k *, any finite Galois extension* L/K *with* $\Gamma := \text{Gal}(L/K)$ *and any* $[B] \in Br(L)^\Gamma$ *the embedding* $RN_{B/L}(B^*) \subseteq L^*$ *induces an isomorphism*

$$H^{-1}(\Gamma, RN_{B/L}(B^*)) \simeq H^{-1}(\Gamma, L^*).$$

We shall see in a moment that the class of reasonable fields is not empty.

Theorem 3. *Let* k *be a reasonable field and* $[D] \in Br(k)$, *then* $SK_1(D) = \{1\}$.

Theorem 4. *Let* $[D] \in Br(k)$, *then* $SK_1(D)$ *is an abelian torsion group of fixed exponent dividing* $i(D)/i_0(D)$ *where* $i_0(D)$ *denotes the greatest squarefree divisor of* $i(D)$.

Corollary 4. *Let* $[D] \in Br(k)$ *be such that* $i(D)$ *is squarefree (i.e. a product of distinct primes), then* $SK_1(D) = \{1\}$. □

Note that Theorem 4 improves the first claim of Lemma 2 and is best possible (cf. (7) in the following §24.); Corollary 4 generalizes parts of Theorem 1.

Proof. We will prove Theorems 3/4 simultaneously: first of all, thanks to Corollary 1 it suffices to study the case

(13) D a k-skew field of index $i(D) = p^f$ for some prime p.

Let $d \in D^*$ be given and assume $RN_{D/k}(d) = 1$; we must show

(14) d (resp. $d^{p^{f-1}}$) $\in [D^*, D^*]$ in case of Theorem 3 (resp. 4).

Now let M be a maximal commutative subfield of D such that $d \in M^*$ (then $|M:k| = p^f$ because of Theorem 4 in §7.); in view of Lemma 7 it suffices to discuss the case "M/k is not purely inseparable" only! If then S denotes the separable closure of k in M we see that S/k is a separable field extension of degree p^e where $1 \leq e \leq f$.

Now let F be a Galois closure of S over k, $\Omega := Gal(F/k)$, Ω_p a p-Sylow subgroup of Ω and $K := Fix_F(\Omega_p)$ the corresponding p-Sylow subfield; then $|K:k|$ and p^f are coprime and hence $A := D \otimes_k K$ remains a skew field - i.e. $i(A) = p^f$, $[A] \in Br(K)$ - (cf. Corollary 8 in §9.) and we have an embedding $SK_1(D) \hookrightarrow SK_1(A)$ (cf. Lemma 3) via the usual embedding $D \to D \otimes_k K$. Therefore, writing $a := d \otimes 1 \in A^*$, in order to prove (14) it will suffice to prove

(15) a (resp. $a^{p^{f-1}}$) $\in [A^*, A^*]$ in case of Theorem 3 (resp. 4).

Now consider $N := M \otimes_k K \subseteq D \otimes_k K = A$ and $T := S \otimes_k K \subseteq N$; both N and T are fields (for otherwise the skew field A would have zerodivisors $\neq 0$ (cf. Exercise 1 in §5. and Exercise 2 in §3.)) such that $|N:K| = p^f$ (i.e. N is maximal commutative in A) and T/K is separable of degree p^e ($1 \leq e \leq f$). Therefore - by the elementary theory of p-groups - we may find an intermediate field L of the extension T/K such that

L/K is *cyclic* of degree p with Galois group (say) Γ.
Consequently (cf. (10) and (8) in §7.)

$$a \in N \subseteq Z_A(N) \subseteq Z_A(L) =: B$$

where B is an L-skew field of index $i(B) = p^{f-1}$ (cf. Lemma 3/Centralizer Theorem in §7.). From all this we conclude (with the notations of Theorem 2): in order to prove (15) it suffices to show

(16) $SK_1(A)|_B = \{1\}$ (resp. $\exp(SK_1(A)|_B) | p^{f-1}$)

Now we use induction on f (f as introduced in (13)): the case "f=1" is then clear from Lemma 7 (L is then maximal commutative in A for degree reasons and we get directly (15) from Lemma 7 ; note that Theorems 3/4 coincide in case f = 1). Now assume f > 1 ; in case of Theorem 4 we get (16) from Corollary 3 because we may assume (by induction hypothesis) that the exponent of $SK_1(B)$ divides p^{t-2} ; finally, in case of Theorem 3 , we conclude (16) from Corollary 2 because we may assume $SK_1(B) = \{1\}$ (again by induction hypothesis) as well as

$$H^{-1}(\Gamma, RN_{B/L}(B^*)) \simeq H^{-1}(\Gamma, L^*) = \{1\}$$

(thanks to (11) and the fact that k is assumed to be reasonable). □

Although the above proof appears first in P. Draxl [1977] its core may be found in disguise in В.И. Янчевский [1975]; the general idea, however, - as well as the general idea of practically everything else in this paragraph - goes back to Wang Shianghaw [1950].

Now how about the mysterious *reasonable fields* ? We have just seen that they are defined in precisely the way which enables us to substantiate Wang Shianghaw's original ideas for a proof of " $SK_1 = \{1\}$ ". Since it is known from Number Theory that $SH^0(A) = \{1\}$ for all central simple K-algebras over *local* fields K (cf. §18. for the definition of a local field; cf. e.g. Prop. 6, §2. in Ch. X of A. Weil [1967] for the above claim) we see:

(17)   Any local field is reasonable.

Consequently Theorem 3 includes a proof of (1) in the preface of Part III. On the other hand one can show (see Satz 1 in P. Draxl [1977]) that also *global* fields are reasonable (cf. §18. for the definition of a global field); this result is much deeper than the corresponding local version: it depends mainly on the description of $SH^0(A)$ over global fields via "Eichler's Norm Theorem" - see also the "Hasse-Schilling-Maaß Theorem" -

(cf. e.g. Prop. 3, §3. in Ch. XI of A. Weil [1967]). We repeat:

(18)   Any global field is reasonable.

Consequently Theorem 3 includes also a proof of (2) in the preface of Part III., usually called "Wang's Theorem" . Note that for the proof of "Eichler's Norm Theorem" the Grunwald-Wang Theorem is not needed (see M. Eichler's own proof in A. Weil *loc. cit.*), although complicated things from global Class Field Theory are still involved.

Summarizing we get from all of the above:

(19) $K_1(H) \simeq R^*_{>0}$ (cf. Theorem 1 and Exercise 3 in §22.) ;

(20) $K_1(D) \simeq K^*$ if D is a K-skew field over a local field K ;

(21) $K_1(D) \simeq RN_{D/K}(D^*)$ if D is a K-skew field and K is global.

Exercise 1 . Give a *quantitative* version of Theorems 3/4 (just as Lemma 4 is a quantitative version of Lemma 2); cf. P. Draxl [1980] .

Exercise 2 . Consider the situation of Theorem 2 : if we denote by r the degree of L/K , then $M_r(B) \simeq A \otimes_K L$ (cf. *(v)* in the Centralizer Theorem in §7.); show

$$K_1(N_{L/K}) = fK_1(\det) \text{ where } \det: GL_r(B) \to K_1(B) .$$

Again f: $K_1(B) \to K_1(A)$ is induced by the embedding.

§ 24 . $SK_1(D) \neq 1$ FOR SUITABLE D

Until 1975 no example for $SK_1(D) \neq \{1\}$ for a suitable skew field D was known; the problem of finding one (or alternatively proving that there is none) often was referred to as the *Tannaka-Artin Problem* (particularly in the Russian literature).

Looking back all this seems strange since we shall introduce later in this paragraph such an example which - in order to be understood - requires hardly more knowledge on skew fields than displayed already in §1. !

The *first example* for $SK_1(D) \neq \{1\}$, however, was somewhat more complicated; it was given by В.П. Платонов [1975]. Further information on that V.P. Platonov developed a whole theory of examples and published it in many (mostly short) papers) and related points of view (partially due to the author) may be obtained from the report P. Draxl & M. Kneser [1980] and the literature list therein (cf. also the remarks at the end of this paragraph).

Now for the just mentioned example: we start with the obvious

Lemma 1 . *If in* Definition 1 in §1. *the field* L *is assumed to be only a skew field (and not a commutative field), then* Definition 1 in §1. *still makes sense and* Lemma 3 in §1. *remains correct.* □

In this context, of course, Lemma 4 in §1. has to be modified since it does not make sense as it stands if L is not commutative; since any automorphism σ of a skew field L preserves the centre (cf. (5) and (12) in §7.) we may consider

$$\sigma_0 := \sigma\big|_{Z(L)} \quad \text{and denote} \quad K_0 := \text{Fix}_{Z(L)}(\sigma_0) .$$

Doing so we note that our new notation coincides with the one from §1. as soon as L is commutative. Copying the proof of Lemma 4 in §1. with the notational changes just introduced we obtain the important

Lemma 2. *Let* L *be a skew field and* $D := L((T;\sigma))$ *(see Lemma 1). If* σ_0 *has infinite order, then* $Z(D) = K_0$, *hence* $|D:Z(D)| = \infty$; *if* σ_0 *has the finite order* n *in* $\mathrm{Aut}(Z(L))$, *then* $Z(D) = K_0((T^n))$, *hence* $|D:Z(D)| = n^2|L:Z(L)|$. □

Now select a commutative field k of characteristic $\neq 2$ and define (skew) fields and automorphisms thereof as follows (in the sense of Lemma 1 above):

(1)
$$L := k((t_1)) \text{ and } \sigma \text{ such that } t_1 \mapsto -t_1, \; \sigma|_k = \mathrm{id}_k;$$
$$E := L((t_2;\sigma));$$
$$F := E((t_3)) \text{ and } \tau \text{ such that } t_3 \mapsto -t_3, \; \tau|_E = \mathrm{id}_E;$$
$$D := F((t_4;\tau)).$$

Thus D is the k-space of iterated formal Laurent series in four variables t_1, t_2, t_3, t_4; we use multiindex notation t^i, $i = (i_4, i_3, i_2, i_1)$, for the monomials $t_1^{i_1} t_2^{i_2} t_3^{i_3} t_4^{i_4}$ and order the quadruples i lexixographically. Thanks to the commutation relations $t_2 t_1 = -t_1 t_2$, $t_4 t_3 = -t_3 t_4$ and $t_j t_i = t_i t_j$ ($i \in \{1,2\}$, $j \in \{3,4\}$) every element $d \in D$ can be uniquely written in the form
$$d = \sum_i a_i t^i \quad (a_i \in k).$$

Set $v(d) := \min\{ i \mid a_i \neq 0 \}$; then $v(dd') = v(d) + v(d')$ (in fact v is even a Henselian valuation of rank 4). Now define (write $T_i := t_i^2$)

(2) $\quad K := k((T_1))((T_2))((T_3))((T_4)) \subseteq D$.

Theorem 1. *Let* D *be according to* (1); *then* D *is a* K*-skew field (* K *according to* (2) *) of index* $i(D) = 4$ *and exponent* $o(D) = 2$; *moreover*

(3) $\quad D \simeq \left(\dfrac{T_1, T_2}{K}\right) \otimes_K \left(\dfrac{T_3, T_4}{K}\right)$ *(in the notation of §14.)*

and

(4) $\quad [D^*, D^*] \subseteq \{ \pm 1 + \sum_{i > 0} a_i t^i \}$.

Furthermore, if k *contains a primitive* 4*-th root of unity* ζ, *then* $\mathrm{RN}_{D/K}(\zeta) = 1$ *but* $\zeta \notin [D^*, D^*]$, *hence* $\mathrm{SK}_1(D) \neq \{1\}$.

Proof. Repeated application of Lemma 1 shows that D is a skew field. Now we use Lemma 2 repeatedly and get from (1) and (2):
$$\mathrm{Fix}_L(\sigma) = k((t_1^2)), \text{ hence } Z(E) = k((t_1^2))((t_2^2)) \text{ and}$$
$$|E:Z(E)| = 4;$$

$Z(F) = Z(E)((t_3))$, hence $\tau_0 := \tau|_{Z(F)}: t_3 \mapsto -t_3$, $\tau_0|_{Z(E)} = \mathrm{id}_{Z(E)}$;

$\mathrm{Fix}_{Z(F)}(\tau_0) = Z(E)((t_3^2)) = k((t_1^2))((t_2^2))((t_3^2))$, hence $Z(D) = (\mathrm{Fix}_{Z(F)}(\tau_0))((t_4^2)) = K$ and $|D:K| = 4|F:Z(F)| = 4|E:Z(E)| = 16$, hence $i(D) = 4$.

The isomorphism (3) and hence $o(D) = 2$ (cf. §9.) is clear enough since evidently we have the two K-algebra isomorphisms (cf. Theorem 2 in §14.)

$$f: A := \left(\frac{T_1, T_2}{K}\right) \xrightarrow{\sim} K \oplus Kt_1 \oplus Kt_2 \oplus Kt_1 t_2$$

and

$$g: B := \left(\frac{T_3, T_4}{K}\right) \xrightarrow{\sim} K \oplus Kt_3 \oplus Kt_4 \oplus Kt_3 t_4$$

which are such that $f(a)f(b) = f(b)f(a)$ for all $a \in A$, $b \in B$. This gives a K-algebra homomorphism $A \otimes_K B \to D$ (cf. Theorem 2 in §5.) which is an isomorphism for dimensional reasons (note that $A \otimes_K B$ is simple). Finally we must study the shape of a commutator in D^* : because of Lemma 1 and the commutation rules of the t_i the construction of the inverse described in the proof of Lemma 3 in §1. goes through with multiindices *mutatis mutandis*. We have $v(dd'd^{-1}d'^{-1}) = 0$ and thus $dd'd^{-1}d'^{-1} = a + \sum_{i>0} a_i t^i$ ($a \in k^*$) where we find $a = \pm 1$ in view of the commutation rules of the t_i. Therefore obviously $\zeta \notin [D^*, D^*]$ but $RN_{D/K}(\zeta) = \zeta^{i(D)} = \zeta^4 = 1$ thanks to (8) in §22. □

Again we point out that Theorem 1 is founded on the (comparatively elementary) results of §1.; save for the definition of the *reduced norm* and its property (8) in §22. nothing about K-skew fields is needed !

It should be remarked that the construction (1) is related to В.А. Липницкий [1976], moreover, the reader should compare the skew field D according to (1) with the skew fields on pp.94,..,104 of the lecture notes N. Jacobson [1975].

We close this paragraph by stating some more interesting results (without proof):

(5) There is a field K such that if [A] runs through $Br(K)$, then $SK_1(A)$ runs through all finite abelian groups (cf. P. Draxl [1975/76] together with Satz 9 in P. Draxl [1977]).

(6) $SK_1(A)$ may be any countable infinite abelian torsion group with given fixed finite exponent for suitable A (cf. the seminar report H.-G. Gräbe [1980]).

There is a field K with the following properties:

(7)
- (i) $SK_1(A) \simeq Z/\frac{i(A)}{o(A)}Z$ for all $[A] \in Br(K)$;
- (ii) given $m, n \in N$ with the same prime factors and such that $m | n$, then there is an $[A] \in Br(K)$ such that $i(A) = n$ and $o(A) = m$.

(cf. P. Draxl [a] ; notation as in §9.)

The last result is related to Ю.Л. Ершов [1982].

As far as applications of "$SK_1(D) \neq 1$" are concerned we refer the reader to В.П. Платонов [1977],[1980], to the book В.Е. Воскресенский [1977] and to the surveys J. Tits [1976/77] and B. Weisfeiler [1982].

Exercise 1 . Consider the skew fields described on pp.94,..,104 of N. Jacobson [1975] (cf. Exercise 1 in §11.) and show that for any such skew field D there is an exact sequence

$$SK_1(D) \longrightarrow Z/\frac{i(D)}{o(D)}Z \longrightarrow 0 \quad \text{(cf. also P. Draxl [a]).}$$

§ 25. REMARKS ON $USK_1(D,I)$

Let A be a central simple K-algebra which admits an involution I of the *second kind*, i.e. a K/k-involution for some separable quadratic subfield k, $\Gamma := \text{Gal}(K/k) = \{\sigma,\text{id}\}$ (cf. §16.). If $S_I(A)$ denotes as usual the set of I-symmetric elements of A we define

(1) $\Sigma_I(A) := \langle S_I(A) \cap A^* \rangle \leq A^*$

the subgroup of A^* which is generated by I-symmetric elements.

Lemma 1. $\Sigma_I(A) \trianglelefteq A^*$.

Proof. $^I x = x$ implies $txt^{-1} = (tx^It)(^I(t^{-1})t^{-1}) \in \Sigma_I(A)$ for all $t \in A^*$. □

Lemma 2. *Given* $a \in A^*$ *such that* I_a *is also an involution, then there exists* $c \in S_I(A) \cap A^*$ *such that* $I_a = I_c$. *Conversely* I_a *is an involution for all* $a \in S_I(A) \cap A^*$.

Proof. By Scharlau's Lemma in §16. I_a is an involution if and only if $a(^I a)^{-1} =: b_a \in K^*$. It follows $N_{K/k}(b_a) = b_a I_a(b_a) = 1$, hence

$a(^I a)^{-1} = {}^\sigma d d^{-1} = {}^I d d^{-1}$ for some $d \in K^*$

by Hilbert's "Satz 90" in §6. which means that we can take $c := ad \in S_I(A) \cap A^*$. The converse is obvious. □

Lemma 3. $\Sigma_I(A) = \Sigma_{I_a}(A)$ *for all* $a \in A^*$, *hence (by Lemma 1 in §16.)* $\Sigma_I(A)$ *depends only on the extension* K/k.

Proof. Thanks to Lemma 2 we may assume $^I a = a$, hence $^I x = x$ implies $x = (xa^{-1})a$ where $xa^{-1}, a \in S_{I_a}(A)$. Therefore $\Sigma_I(A) \subseteq \Sigma_{I_a}(A)$ and even "=" for reasons of symmetry. □

A very important result is (cf. Exercise 1)

Vaserstein's Lemma. *Let* D *be a K-skew field which admits an involution* I *of the second kind, then* $[D^*,D^*] \subseteq \Sigma_I(D)$. □

Now let A be a central simple K-algebra which admits an involution I of the second kind (note that this implies $K \neq F_2, F_3$) and D its skew field component - i.e. we may assume $A = M_n(D)$ -, then Lemma 2 in §16. and Lemma 3 imply that for the study of $\Sigma_I(A)$ we may also assume
$$^I(d_{ij}) = (^I d_{ji}) \text{ in } M_n(D) = A.$$
We want to show (cf. Theorem 4 in §20.) $SL_n(D) = [A^*, A^*] \subseteq \Sigma_I(A)$ if $n \geq 2$ (the case $n = 1$ is covered by Vaserstein's Lemma). Indeed, if the latter were false, Lemma 1 would imply that $\Sigma_I(A) \cap SL_n(D)$ is a proper normal subgroup of $SL_n(D)$, hence (see Theorem 2 in §21.)
$$\Sigma_I(A) \cap SL_n(D) \subseteq Z(SL_n(D)).$$
The latter, however, leads to a contradiction (see Exercise 2), hence

Jančevskiĭ's Lemma. *If A is a central simple K-algebra with an involution I of the second kind, then* $[A^*, A^*] \subseteq \Sigma_I(A)$. □

Definition 1. *Let A be a central simple K-algebra with an involution I of the second kind, then*
$$UK_1(A, I) := A^* / \Sigma_I(A)$$
is called the "unitary Whitehead group of A and I".

From what we have seen above $UK_1(A, I)$ is always abelian (and hence a homomorphic image of $K_1(A)$) and depends only on the extension K/k where $k = K \cap S_I(A)$; in particular: $UK_1(K, I) = K^*/k^*$. Moreover, Lemma 5 in §22. yields $RN_{A/K}(\Sigma_I(A)) \subseteq k^*$, hence $RN_{A/K}$ induces a homomorphism
(2) $UK_1(I, RN_{A/K}): UK_1(A, I) \longrightarrow UK_1(K, I) = K^*/k^*$.

Definition 2. *Let A be a central simple K-algebra with an involution I of the second kind, then*
$$USK_1(A, I) := \text{Ker } UK_1(I, RN_{A/K}) =$$
$$= \{ a \in A \mid RN_{A/K}(a) \in k^* \} / \Sigma_I(A)$$
is called the "reduced unitary Whitehead group of A and I".

Now, by modifying the methods from §§23/25 appropriately one can establish a theory of USK_1 in formally the same way as in the case SK_1. Most of this is due to V.I. Jančevskiĭ (cf. the survey P. Draxl [1979] and the references therein). For different aspects see P. Draxl [1980],[1982].

Exercise 1. Prove Vaserstein's Lemma (see В.П. Платонов & В.И. Янчевский [1973]).

Exercise 2. Complete the proof of Jančevskiĭ's Lemma (see В.И. Янчевский [1974]).

BIBLIOGRAPHY

I.T. Adamson

[1954]   Cohomology Theory for Non-Normal Subgroups and Non-Normal Fields, *Proc. Glasgow Math. Assoc.* 2 (1954/56) 66-76

A.A. Albert

[1939]   Structure of Algebras, *AMS Coll. Publ.* 24, New York (1939)

[1972]   Tensor Products of Quaternion Algebras, *Proc. AMS* 35 (1972) 65-66

S.A. Amitsur

[1962]   Homology Groups and Double Complexes for Arbitrary Fields, *J. Math. Soc. Japan* 14 (1962) 1-25

E. Artin

[1957]   Geometric Algebra, *Interscience Publ.*, New York (1957)

E. Artin, C.J. Nesbitt, R.M. Thrall

[1948]   Rings with Minimum Condition, *Univ. of Michigan Press*, Ann Arbor (1948)

E. Artin, J. Tate

[1968]   Class Field Theory, *Benjamin*, New York & Amsterdam (1968)

E. Artin, G. Whaples

[1943]   The Theory of Simple Rings, *Amer. J. Math.* 65 (1943) 87-107

A. Babakhanian

[1972]   Cohomological Methods in Group Theory, *Marcel Dekker*, New York (1972)

R. Baeza

[1978]   Quadratic Forms over Semilocal Rings, *Springer LNM* 655, Berlin, Heidelberg, New York (1978)

N. Bourbaki

[1958]   Algèbre, Ch. 8, Modules et anneaux semisimples, *Hermann*, Paris (1958)

J.W.S. Cassels, A. Fröhlich (editors)

[1967]   Algebraic Number Theory, *Academic Press*, New York & London (1967)

P.M. Cohn

[1971] Free Rings and their Relations, *Academic Press*, New York & London (1971)

[1974/77] Algebra (2 volumes), *Wiley*, New York (1974 & 1977)

[1977] Skew Field Constructions, *LMS Lecture Note Series* 27, Cambridge (1977)

C.W. Curtis, I. Reiner

[1962] Representation Theory of Finite Groups and Associative Algebras, *Interscience Publ.*, New York (1962)

F. DeMeyer, E. Ingraham

[1971] Separable Algebras over Commutative Rings, *Springer LNM* 181, Berlin, Heidelberg, New York (1971)

M. Deuring

[1935] Algebren, *Springer*, Berlin, Heidelberg (1935)

J. Dieudonné

[1963] La géométrie des groupes classiques, *Springer*, Berlin, Heidelberg (1963)

P. Draxl

[1975] Über gemeinsame separabel-quadratische Zerfällungskörper von Quaternionenalgebren, *Nachr. Akad. Wiss. Göttingen* 16 (1975)

[1975/76] Corps gauches dont le groupe des commutateurs n'est pas égal au noyau de la norme réduite, *Séminaire Delange-Pisot-Poitou* 17 (№ 26) (1975/76)

[1977] SK_1 von Algebren über vollständig diskret bewerteten Körpern und Galoiskohomologie abelscher Körpererweiterungen, *J. reine u. angew. Math.* 293/294 (1977) 116-142

[1979] Corps gauches à involution de deuxième espèce, *SMF Astérisque* 61 (1979) 63-72

[1980] Eine Liftung der Dieudonné-Determinante und Anwendungen die multiplikative Gruppe eines Schiefkörpers betreffend, *in: Teil II of* P. Draxl & M. Kneser [1980], 101-116

[1982] Normen in Diedererweiterungen von Zahlkörpern, *Abh. Wiss. Ges. Braunschweig* 33 *(Festband der Dedekind-Tagung 1981)* 99-116

[a] On Tame Skew Fields, *in preparation*

P. Draxl, M. Kneser (editors)

[1980] SK_1 von Schiefkörpern, *Springer LNM* 778, Berlin, Heidelberg, New York (1980)

S. Eilenberg, S. MacLane

[1948] Cohomology and Galois Theory. I. Normality of Algebras and Teichmüller's Cocycle, *Trans. AMS* 64 (1948) 1-20

L. Elsner

[1979] On Some Algebraic Problems in Connection with General Eigenvalue Algorithms, *Lin. Alg. and Applic.* 26 (1979) 123-138

Ю.Л. Ершов (Yu.L. Eršov)

[1982] Гензелевы нормирования тел и группа SK_1 , Мат. Сборник 117 (1982) 60-68

B. Fein, M. Schacher

[1976] The Ordinary Quaternions over a Pythagorean Field, *Proc. AMS* 60 (1976) 16-18

[1982] Relative Brauer Groups III, *J. reine u. angew. Math.* 335 (1982)

B. Fein, W.M. Kantor, M. Schacher

[1981] Relative Brauer Groups II, *J. reine u. angew. Math.* 328 (1981) 39-57

H.-G. Gräbe

[1980] Über die Umkehraufgabe der reduzierten K-Theorie, *Seminar Eisenbud-Singh-Vogel* 1, *Teubner-Texte Math.* 29 (1980) 94-107

G. Hochschild

[1955] Restricted Lie Algebras and Simple Associative Algebras of Characteristic p , *Trans. AMS* 80 (1955) 135-147

N. Jacobson

[1975] *PI*-Algebras, An Introduction, *Springer LNM* 441, Berlin, Heidelberg, New York (1975)

В.И. Янчевский (V.I. Jančevskiĭ)

[1974] Простые алгебры с инволюциями и унитарные группы, Мат. Сборник 93 (1974) 368-380
= Simple Algebras with Involution, and Unitary Groups, *Math. USSR Sbornik* 22 (1974) 372-385

[1975] Коммутанты простых алгебр с сюръективной приведенной нормой ДАН СССР 221 (1975) 1056-1058
= The Commutator Subgroups of Simple Algebras with Surjective Reduced Norms, *Soviet Math. Dokl.* 16 (1975) 492-495

M. Kervaire, M. Ojanguren (editors)

[1981] Groupe de Brauer, *Springer LNM* 844, Berlin, Heidelberg, New York (1981)

F. Klein

[1926/27] Vorlesungen über die Entwicklung der Mathematik im 19. Jahrhundert (2 volumes), *Springer*, Berlin (1926 & 1927)

M.-A. Knus, M. Ojanguren

[1974] Théorie de la Descente et Algèbres d'Azumaya, *Springer LNM* 389, Berlin, Heidelberg, New York (1974)

В.В. Курсов (V.V. Kursov)

[1979] О коммутанте полной линейной группы над телом, ДАН БССР 23 (1979) 869-871

T.Y. Lam

[1973] The Algebraic Theory of Quadratic Forms, *Benjamin*, Reading (1973)

В.А. Липницкий (V.A. Lipnickiĭ)

[1976] К проблеме Таннака-Артина над специальными полями, ДАН СССР 228 (1976) 26-29
= On the Tannaka-Artin Problem over Special Fields, *Soviet Math. Dokl.* 17 (1976) 639-642

Y. Matsushima, T. Nakayama

[1943] Über die multiplikative Gruppe einer p-adischen Divisionsalgebra, *Proc. Imp. Acad. Japan* 19 (1943) 622-628

А.С. Меркурев (A.S. Merkur'ev)

[1981] О символе норменного вычета степени 2, ДАН СССР 261 (1981) 542-547
= On the Norm Residue Symbol of Degree 2, *Soviet Math. Dokl.* 24 (1981) 546-551

А.С. Меркурев, А.А. Суслин (A.S. Merkur'ev, A.A. Suslin)

[a] *to appear in* ДАН СССР (presumedly in 1982)

[b] K-Cohomology of Severi-Brauer Varieties and Norm Residue Homomorphism, *Preprint* (1981)

J. Milnor

[1971] Introduction to Algebraic K-Theory, *Ann. of Math. Studies* 72, Princeton (1971)

O.T. O'Meara

[1963] Introduction to Quadratic Forms, *Springer*, Berlin, Heidelberg (1963)

В.П. Платонов (V.P. Platonov)

[1975] О проблеме Таннака-Артина, ДАН СССР 221 (1975) 1038-1041
= On the Tannaka-Artin Problem, *Soviet Math. Dokl.* 16 (1975) 468-473

[1977] Бирациональные свойства приведенной группы Уайтхеда, ДАН БССР 21 (1977) 197-198

[1980] Algebraic Groups and Reduced K-Theory, *Proc. ICM (Helsinki, 1978)*, Helsinki (1980) 311-317

В.П. Платонов, В.И. Янчевский (V.P. Platonov, V.I. Jančevskiĭ)

[1973] Структура унитарных групп и коммутант простой алгебры над глобальными полями, ДАН СССР 208 (1973) 541-544
= The Structure of Unitary Groups and the Commutator Group of a Simple Algebra over Global Fields, *Soviet Math. Dokl.* 14 (1973) 132-136

C. Procesi

[1973] Rings with Polynomial Identities, *Marcel Dekker*, New York (1973)

U. Rehmann

[1978] Zentrale Erweiterungen der speziellen linearen Gruppen eines Schiefkörpers, *J. reine u. angew. Math.* 301 (1978) 77-104

[1980] Kommutatoren in $GL_n(D)$, *in: Teil II of* P. Draxl & M. Kneser [1980], 117-123

I. Reiner

[1975] Maximal Orders, *Academic Press*, New York & London (1975)

C. Riehm

[1970]   The Corestriction of Algebraic Structures, *Inventiones math.*
          11 (1970) 73-98

S. Rosset

[1977]   Abelian Splitting of Division Algebras of Prime Degree, *Comment.
          Math. Helvetici* 52 (1977) 519-523

[a]   The Corestriction Preserves Sums of Symbols, *Preprint*

W. Scharlau

[1969]   Über die Brauer-Gruppe eines Hensel-Körpers, *Abh. Math. Sem.
          Univ. Hamburg* 33 (1969) 243-249

[1975]   Zur Existenz von Involutionen auf einfachen Algebren, *Math. Z.*
          145 (1975) 29-32

[1981]   Zur Existenz von Involutionen auf einfachen Algebren II,
          Math. Z. 176 (1981) 399-404

O.F.G. Schilling

[1950]   The Theory of Valuations, *AMS Math. Surveys* 4, New York (1950)

J.-P. Serre

[1962]   Corps Locaux, *Hermann*, Paris (1962)

S.S. Shatz

[1972]   Profinite Groups, Arithmetic, and Geometry, *Ann. of Math.
          Studies* 67, Princeton (1972)

R.L. Snider

[1979]   Is the Brauer Group Generated by Cyclic Algebras ? *in:*
          D. Handelman & J. Lawrence, Ring Theory, Waterloo 1978, *Springer
          LNM* 734, Berlin, Heidelberg, New York (1979), 280-301

J. Tate

[1976]   Relations between K_2 and Galois Cohomology, *Inventiones math.*
          36 (1976) 257-274

O. Teichmüller

[1936]   p-Algebren, *Deutsche Math.* 1 (1936) 362-388

[1940]   Über die sogenannte nichtkommutative Galoissche Theorie und die
          Relation $\xi_{\lambda,\mu,\nu}\xi_{\lambda,\mu\nu,\pi}\xi^{\lambda}_{\mu,\nu,\pi} = \xi_{\lambda,\mu,\nu\pi}\xi_{\lambda\mu,\nu,\pi}$, *Deutsche Math.*
          5 (1940) 138-149

R.C. Thompson

[1961]   Commutators in the Special and General Linear Group, *Trans.
          AMS* 101 (1961) 16-33

J.-P. Tignol

[1981]   Corps à involution neutralisés par une extension abélienne
          élémentaire, *in:* M. Kervaire & M. Ojanguren [1981], 1-34

J. Tits

[1976/77] Groupes de Whitehead de groupes algébriques simples sur un
          corps, *Séminaire Bourbaki* 29 (№ 505) (1976/77)

В.Е. Воскресенский (V.E. Voskresenskiĭ)

[1977]   Алгебраические Торы, Наука, Москва (1977)

Wang Shianghaw

[1950]   On the Commutator Group of a Simple Algebra, *Amer. J. Math.* **72** (1950) 323-334

W.C. Waterhouse

[1982]   Linear Maps Preserving Reduced Norms, *Lin. Alg. and Applic.* **43** (1982) 197-200

A. Weil

[1967]   Basic Number Theory, *Springer*, Berlin, Heidelberg, New York (1967)

B. Weisfeiler

[1982]   Abstract Homomorphisms between Subgroups of Algebraic Groups, *Notre Dame University Lecture Notes Series*, (1982)

E. Weiss

[1969]   Cohomology of Groups, *Academic Press*, New York & London (1969)

E. Witt

[1936]   Konstruktion von galoisschen Körpern der Charakteristik p zu vorgegebener Gruppe der Ordnung p^f, *J. reine u. angew. Math.* **174** (1936) 237-245

[1937]   Zyklische Körper und Algebren der Charakteristik p vom Grad p^n, *J. reine u. angew. Math.* **176** (1937) 126-140

K. Yoshida

[1965]   Functional Analysis, *Springer*, Berlin, Heidelberg, New York (1965)

D. Zelinsky (editor)

[1976]   Brauer Groups, *Springer LNM* **549**, Berlin, Heidelberg, New York (1976)

THESAURUS

M. Deuring [1935]	A.A. Albert [1939]	E. Artin et al. [1948]	these notes [1982]
A/K einfache normale Algebra	A normal simple algebra over K	A simple algebra with center K	A central simple K-algebra
A^*	A^{-1}	A^*	A^{op}
K_n	M_n, M	M_n	$M_n(K)$
$(A:K)$	order of A	$(A:K)$ = degree of A	$|A:K|$ = degree of A
$\sqrt{(A:K)}$	degree of A	$\sqrt{(A:K)}$	reduced degree of A
–	A^B = A-commutator of B	A^B = commutator of B	$Z_A(B)$ = centralizer of B in A
direktes Produkt A×B	direct product A×B	Kronecker product $A \times_K B$	tensor product $A \otimes_K B$
A_L	A_L	$A \times_K L$	$A \otimes_K L$
Brauersche Gruppe	class group	–	Brauer group Br(K)
{A} (in multiplicative notation)	(A) (in multiplicative notation)	–	[A] (in additive notation)
Index von A	index of A	–	i(A) = index of A
Exponent von A	exponent of A	exponent of A	o(A) = exponent of A
verschränktes Produkt (x,L)	crossed product (L,x)	crossed product $(L/K, x_{\sigma,\tau})$	crossed product (x,L/K)
zyklische Algebra (a,L,σ)	cyclic algebra (L,σ,a)	cyclic algebra (L/K,σ,a)	cyclic algebra (a,L/K,σ)

INDEX

Adamson cohomology group	122	$C^1(\Gamma,M)$	34
Albert's Criterion	87	$C^2(\Gamma,M)$	93
Albert's Main Theorem	110	centralizer	40
algebra	25	Centralizer Theorem	42
Clifford -	105	central simple algebra	29
cyclic -	49	Clifford algebra	105
Galois -	37	conjugation (in cohomology)	100
p- -	106	corestriction (in cohomology)	101
PI- -	122	corestriction (of algebras)	53
quaternion -	78, 84	crossed product	94
algebra homomorphism	25	Crossed Product Theorem	96
Alkohol	1	cup product	96
alternating group	142	cyclic algebra	49
Amitsur cohomology group	122		
antiautomorphism	112	Dedekind's Lemma	33
Artinian module	8	Dickson's Theorem	91
Artin-Schreier Theorem	63	Dieudonné determinant	135
Artin's Lemma	33	dihedral group	142
Artin-Whaples Theorem	28		
		Eichler's Norm Theorem	165
$B^1(\Gamma,M)$	34	elementary matrices	8
$B^2(\Gamma,M)$	93	elementary operations	9
balanced map	18	exponent (of an algebra)	66
bimodule	22		
Brauer group	61	flat module	21
relative -	61	formally real field	105
Bruhat normal form	131	Frobenius' Theorem	73
strict -	128		

Galois algebra	37	maximal commutative subfield	45
Galois module	36	maximal commutative subring	40
Γ-module	33	Mazur-Gelfand Theorem	123
General Merkur'ev-Suslin Theorem	121	Merkur'ev-Suslin Theorem	119
global field	123	General -	121
Grunwald Theorem	125	metabelian field extension	65
Grunwald-Wang Theorem	125	module	8
		Artinian -	8
$H^{-1}(\Gamma,M)$	161	bi -	22
$H^{0}(\Gamma,M)$	34	flat -	21
$H^{1}(\Gamma,M)$	34	Γ- -	33
$H^{2}(\Gamma,M)$	93	Galois -	36
Hasse-Schilling-Maaß Theorem	165	Noetherian -	8
Hilbert's "Satz 90"	35	projective -	16
Hochschild-Serre spectral sequence	62		
Hochschild's Theorem	110	Noetherian module	8
		Noether's Equations	34
ideal	8	norm (of a Γ-module)	33
idempotent	14	norm (of an algebra)	143
index (of an algebra)	59	reduced -	145
index reduction factor	67	norm residue class group	34
inflation (in cohomology)	97	norm residue homomorphism	83
involution (of a ring)	112		
inverse ring	27	opposite ring	27
irreducible module	8	ordinary quaternions	3
		Ostrowski's Second Theorem	123
Jančevskiĭ's Lemma	172		
		p-algebra	106
$K_1(A)$	137	perfect group	140
$K_2(K)$	82	permutation matrix	9
Köthe's Theorem	64	*PI*-algebra	122
		power norm residue algebra	78
Laurent series	5	projective module	16
linear extension	27	Pythagorean field	105
local field	123		
		quadratic form	105
Matsumoto's Theorem	83	quaternion	3

quaternion algebra	78, 84	Tate's Reciprocity Lemma	88
quaternion group	3	tensor product (of modules)	18
		tensor product (of algebras)	25
real quaternions	3	tower formulae	144
reasonable field	163	reduced -	150
reduced degree	59	trace (of an algebra)	143
- norm	145	reduced -	145
- norm residue group	155		
- tower formulae	150	$UK_1(A,I)$	172
- trace	145	unitary Whitehead group	172
- unitary Whitehead group	172	$USK_1(A,I)$	172
- Whitehead group	155		
relative Brauer group	61	Vaserstein's Lemma	171
restriction (in cohomology)	99		
Riehm's Theorem	118	Wang's Theorem	166
Rosset's Theorem	90	Wedderburn's Main Theorem	15
Rosset-Tate Theorem	89	Wedderburn's Theorem	73
		Whitehead group	137
Scharlau's Criterion	116	reduced -	155
Scharlau's Lemma	115	Whitehead Lemma	137
Scharlau's Theorem	116	Witt's Theorem	110
semisimple ring	11		
$SH^0(A)$	155		
similar	60		
simple module	8		
simple ring	11		
$SK_1(A)$	155		
skew field component	59		
K-skew field	29		
Skolem-Noether Theorem	38		
soluble field extension	65		
splitting field	61		
strict Bruhat normal form	128		
supernatural number	67		
symbol (in K-Theory)	83		
Tannaka-Artin Problem	167		

For EU product safety concerns, contact us at Calle de José Abascal, 56–1º, 28003 Madrid, Spain or eugpsr@cambridge.org.